Founders of the Future

Campos Ibéricos
Bucknell Studies in Iberian Literatures and Cultures

Series Editors
Isabel Cuñado, Bucknell University
Jason McCloskey, Bucknell University

Campos Ibéricos is a series of monographs and edited volumes that focuses on the literary and cultural traditions of Spain in all of its rich historical, social, and linguistic diversity. The series provides a space for interdisciplinary and theoretical scholarship exploring the intersections between literature, culture, the arts, and media from medieval to contemporary Iberia. Studies on all authors, texts, and cultural phenomena are welcome and works on understudied writers and genres are specially sought.

Titles in the Series

Óscar Iván Useche, *Founders of the Future: The Science and Industry of Spanish Modernization*
Carrie L. Ruiz and Elena Rodríguez-Guridi, eds., *Shipwreck in the Early Modern Hispanic World*
Joan L. Brown, *Calila: The Later Novels of Carmen Martín Gaite*
Andrés Lema-Hincapié and Conxita Domènech, eds., *Indiscreet Fantasies: Iberian Queer Cinema*
Katie J. Vater, *Between Market and Myth: The Spanish Artist Novel in the Post-Transition, 1992–2014*

Founders of the Future

The Science and Industry of Spanish Modernization

Óscar Iván Useche

Bucknell University | University Press
Lewisburg, Pennsylvania

Library of Congress Cataloging-in-Publication Data

Names: Useche, Óscar Iván, author.
Title: Founders of the future : the science and industry of Spanish modernization / Óscar Iván Useche.
Description: Lewisburg, Pennsylvania : Bucknell University Press, [2022] | Series: Campos ibéricos: Bucknell studies in Iberian literatures and cultures | Includes bibliographical references and index.
Identifiers: LCCN 2021021436 | ISBN 9781684483853 (paperback) | ISBN 9781684483860 (hardback) | ISBN 9781684483877 (epub) | ISBN 9781684483884 (mobi) | ISBN 9781684483891 (pdf)
Subjects: LCSH: Industrialization—Spain—History—19th century. | Industrialization—Spain—History—20th century. | Organizational change—Spain—History—19th century. | Organizational change—Spain—History—20th century.
Classification: LCC HC385 .U77 2022 | DDC 330.946—dc23
LC record available at https://lccn.loc.gov/2021021436

A British Cataloging-in-Publication record for this book is available from the British Library.

www.bucknelluniversitypress.org

Distributed worldwide by Rutgers University Press

Manufactured in the United States of America

For Violetta and Raphaella

Contents

Note on Translations

All translations in *Founders of the Future*, unless otherwise noted, are mine.

Founders of the Future

Introduction

REACHING OUT INTO THE FUTURE

Symbolic Impact

Industrial modernization transformed Europe's social, political, and cultural organization. It not only altered the social fabric, reshaping class structures, mechanisms of capital accumulation and distribution of wealth, and systems of human interaction, but also created new cognitive paradigms, turning into a powerful allegorical engine that provided novel and effective tools for evaluating society and negotiating identity. This process was particularly contentious in Spain, where attempts at consolidating scientific and technological traditions to support industrial growth resulted in irreconcilable tensions that delayed or even diverted progress. Industrialization was not always an answer for reaching out into the future. As a matter of fact, by the end of the nineteenth century a series of conflictive positions marked the cultural agenda with regard to national modernization. On the one hand, industrial progress nurtured in the collective imagination the possibility of overcoming the country's long history of economic and social difficulties. This prospect persuaded progressivist sectors to celebrate science and technology as possible answers to the national backwardness. On the other hand, and despite the clear necessity of social transformation to facilitate innovation and growth, Spaniards placed tradition before their own material aspirations, a factor that conservative circles harnessed to glorify the past and censure modernization as a negative force. Industrial development was thus broadly seen as both an opportunity and a dangerous contingency—it provided the occasion to join European modernity but threatened to debilitate the principles over which national identity had consolidated.

This study proposes a new way of reading *fin-de-siglo* discursive production by uncovering some of the ideas and subjectivities that emerged at the intersection of industrial and national projects. While it assumes that industrialization interacted with the principles of discursive production in a diverse set of social, cultural, and political contexts, the analyses it advances are not concerned with the way in which texts respond to or describe industry and its impact on society. Rather, each chapter explores the productive incorporation of industrial language and imagery into a wide variety of textual formats to offer attentive diagnoses of the country's most urgent problems and challenges. Ranging from the popularization of science to fiction, and including journalism, travel accounts, and political essays, this book shows how the complexities of national modernization were negotiated predominantly through the act of writing.

Cultural engagement with the materiality of industry produced a constellation of reflections about the nature of the social itself. To be sure, one cannot talk about a homogeneous or single way in which industrial transformation altered the perception of reality. On the contrary, subjected to particular beliefs and following specific political agendas, artists, educators, scientists, politicians, journalists, and cultural critics generated a vast array of complementary perspectives on the national situation. Questions such as economic recovery, the consolidation of peripheral nationalisms, the growing unrest of the working classes, or the environmental and human cost of technological progress, to mention just some of the issues studied in the following pages, were addressed from a plurality of critical approaches, often defying the ideological divisions that polarized the country's political life. In fact, the appropriation of industrial referents into the cultural imagination was a versatile and multifaceted process that not only involved certain convictions but also encompassed different forms of knowing.

Philosophy, which profusely permeated discursivity at the turn of the twentieth century, was the backdrop and base for authors' attempts at harmonizing antagonistic views of the country. Thinkers tackled tensions between spirit and matter and faith and reason that had historically complicated the adoption of progressive ideas, scientific discoveries, and technological advances on the peninsula. By formulating sophisticated yet untenable philosophical loopholes, they showed that in order for modernization to be attained, confrontations between tradition and progress needed to be overcome. In general terms, industrial growth was perceived as a negative, profane force whose capacity to alter reality affected the pillars of identity, producing a similar effect to that which Walter Benjamin conceptualizes as a loss of aura. In the same way that mechanical reproduction disrupted our understanding of cultural production, industrial development modified how society was perceived. If Benjamin defines *aura* as a mystical state, a reverential attitude connected to the unrepeatable quality of the artistic artifact,[1] one can extrapolate this articulation to approach the physical and ideological transfiguration of Spain during this period as a loss of aura. A new sense of class stability, disruptions to ideas of center and periphery, and the destruction of nature were indeed some of the aspects that complicated the negotiations of national identity as a bastion of historical continuity with the past.

Rather than focusing on the transformation of society per se, this study centers on the symbolic spaces—narrative strategies, rhetorical devices, metaphorical systems—that emerged in the discursive negotiations of such changes. Even though scholarship on Spanish industrialization has been extensive, few studies have delved into its semiological impact on the form through which national problems were evaluated.[2] Yet it is precisely at the metaphorical and allegorical levels that industrial modernization affected the cultural and social realms, changing how reality was abstracted into a new language that permeated all forms of discourse. Images, concepts, and ideas related to industry became ubiquitous referents in discussions about national consolidation. Mechanization and automation, for instance, stand out as symbols of the country's profound transformation—the particular dynamism, structural organization, and operative conception of modern industries evinced the richness of the past, the mutability of the present, and the possibilities of the future. At the semiotic level, industrial metaphors and allegories offered new platforms (methodological, thematic, and discursive) for interpreting reality. As George Lakoff and Mark Johnson explain, "Metaphor is not merely a matter of language. It is a matter of conceptual structure. And conceptual structure is not merely a matter of the intellect—it involves all the natural dimensions of our experience."[3] Contact with industrial reality, consequently, produced a rich repertoire of conceptual and analytical tools that facilitated the production of coherent discourses about society, education, politics, and the economy.

Founders of the Future's main goal is therefore to identify the dialectical exchange between industrialization and society and to examine the rhetorical devices and narrative strategies that emerged at this interface. In order to show how industry's operational, scientific, and technical principles transformed the cultural premises of textual production, the analyses included here assume that the dialogue between industry and culture takes place over a dynamic network of relationships. Material culture's central proposition is useful in that regard to understand how the world of ideas interacted with the concrete realities of industrial transformation. According to this theoretical approach, objects have an impact on the social world by altering abstract articulations of reality—metaphors and allegorical constructions of industry thereby represented a novel and pertinent form of understanding social change.

Notwithstanding the fact that mechanization in Spain was already propagating at the end of the eighteenth century, this study focuses on its flourishing as a discursive and allegorical referent during the last third of the nineteenth and the first decades of the twentieth centuries. This period, which coincides with the Bourbon restoration, was characterized by a political consensus regarding the need for economic and institutional reforms that favored the consolidation of a strong industrial base in the country. The development of large-scale industries in Spain encompassed diverse sectors, including mining, manufacturing, and transportation. However, the steel industry stood out as the epitome of the possibilities of modernization and the image of progress's dynamism and transformative power. In fact, the symbolic association of the production of metals with a society in flux is rather apposite to this book's proposal. The functioning of the steel industry presented a new means of circulating ideas about time, energy, productivity, and collectivity that reshaped the projects of national modernization. The foundry became a point of reference in operational models of society that authors used to tackle the tensions and contradictions of the period. This convergence between industrial materiality and social diagnosis gave rise to a new interpretative lens, a novel insight into the country's transformation that I have termed *social foundry*, and whose conceptual grounds are expounded in chapter 1.

The social foundry's use of industrial signifiers, practices, and concepts emphasizes three distinctive and essential elements of modernization: human agency, materiality, and causality. In contrast to other scientifically or technically based abstractions of reality, such as social Darwinism or social thermodynamics, the social foundry's consideration of human intervention—in the form of intellectual or physical labor—provided a concrete and relevant referential system for the formulation of plans for social reform. Additionally, in its materiality this frame of reference privileged the heuristic over the conceptual, encompassing practical and empirical elements related to the circulation of industrial objects in society. Finally, the social foundry's logic transcended causality and determinism by considering the multiple directions that society could take according to the administration of its human and natural resources. Attention to the constitution of social foundries, then, offers new perspectives on how economic development, education, social transformation, scientific development, and environmental impact were evaluated in fin-de-siglo Spain as key elements in the forging of the modern nation.

The Future Was Here

The devastating effects of the 2008–2014 financial crisis in Spain has obscured the country's recent achievements in terms of scientific research, technological advancement, and industrial development. In sectors such as renewable energy, medicine, and transportation, Spain's innovations were on a par with those of other European nations.[4] Even as late as 2010 (in the midst of the crisis), the high-speed rail network, an indicator of development, was still in expansion. That year, King Juan Carlos I inaugurated a new line that connected the cities of Madrid and Valencia. In his speech on the occasion, the monarch stressed the importance of continuing to work toward complete connectivity on the peninsula, an effort that would not only strengthen the economic and cultural links among different regions but would also consolidate the idea of a unified national identity: "Esta extraordinaria conexión, concebida para dar el mejor y más rápido servicio a los ciudadanos, va a favorecer sin duda nuestra economía, y nuestra cohesión social y territorial" (This exceptional connection, conceived as a mechanism for providing citizens with a better and faster service, will undoubtedly favor our economy and our social and territorial cohesion).[5] Perhaps not surprisingly, it was these same goals that had mobilized investors to promote the construction of the first railroad line back in 1848. However, what in the past represented a small but significant step in the country's modernization, in the twenty-first century positioned Spain as one of the most advanced nations in terms of railroad infrastructure.[6] Moreover, with this huge leap, Spain demonstrated its total integration into the dynamics of European modernization, something unimaginable 150 years earlier. Looking at the past to explore the elements that enabled this integration thus reveals the complex assimilation of progress into the collective imagination and the extent to which this symbolic transaction shaped cultural, social, and political ideas about the country.

The pinnacle of nineteenth-century industrialization in Spain coincided with a period of relative political stability. The restoration of the monarchy in 1875 and the establishment of the long-lasting conservative political system that ruled the country for more than three decades show indeed how difficult it was to transition from traditional forms of government to progressive models of state administration that were advantageous to the projects of national modernization.[7] The peaceful shifting in power between liberals and conservatives, which was enforced as a corrective measure to the historical confrontation of these irreconcilable ideologies, provided a sense of economic and social improvement whose inefficacy would be later exposed by the moral and military defeat in the 1898 war with the United States. While achieving an ideological integration of the peninsula had proven to be challenging, this loss confirmed the problematic underdevelopment of the country. Yet, at the same time, advances in infrastructure and the expansion of the industrial base showed prospects of economic recovery that were contingent on the collective efforts of various sectors in society. This was one of the aspects that King Juan Carlos remarked on in his 2010 speech during the inauguration of the high-speed line between Madrid and Valencia: "[Este logro] simboliza . . . el resultado feliz de años de trabajo, de férrea voluntad y de generosa unidad" ([This achievement] symbolizes the happy results of years of work, iron will, and generous unity).[8] Science and technology were key in giving shape to a new vision of the country, one that only materialized in the late twentieth century thanks to the relentless labor and generosity of scientists and educators. Their success, however, had not been simply won.

As a matter of fact, promotion and development of science and technology during the nineteenth century were constantly compromised by the idea that progress could impinge on traditional values. Cultural and political agents, in this context, aligned with diverse and occasionally incompatible perspectives when negotiating social changes. Progressivists, for instance, proposed a thorough re-evaluation of the country's political and economic structures in order to respond more effectively to the new dynamics of industry. Conservatives, meanwhile, rejected any form of power redistribution or social mobility, thus questioning the very notion of modernization and limiting its potential. Traditionalist thinkers who aligned with these ideas resisted the reorganization of society, arguing that its structure was essential to keeping national cultural values intact. Yet conservatives and traditionalists did not disdain the material possibilities of progress; on the contrary, most of them participated actively in entrepreneurial initiatives to promote industrial development, making sure that social and political control remained in the hands of the governing elites. Liberals, though, contested this position by exposing and denouncing the social crisis that co-opting the benefits of modernization could generate. In its different chapters, *Founders of the Future* explores this ideological complexity, tracing its protean nature and the repercussions it had for the conceptualization of Spain as a modern nation.

An example may be pertinent to illustrate this point. Consider Ramón Torres Muñoz de Luna's laudatory poem written in 1861 to celebrate the inauguration of the railroad line joining Madrid and León. Exalting modernization, in it the poet draws an interesting comparison between the achievements of queens Isabel I and Isabel II in terms of political and territorial cohesion. Titled "Las dos Isabeles" (The two Isabels), the panegyric looks at the past to celebrate the monarchy's and the Church's crucial role in forging the national identity. The piece's theme is in this regard paradigmatic of the problematic assimilation of industrial, scientific, and technological progress in Spain. When referring to Isabel I, for example, the poet underlines, "Un mundo mereciste a Dios clemente/Que a Colón señaló tu diestra mano,/Cual precioso diamante refulgente/Perdido entre las olas del océano" (One world you deserved from merciful God/the one your skillful hand showed to Columbus,/like a brilliant diamond/lost between the ocean waves).[9] Similarly, the second Isabel knew how to appreciate the possibilities of the unknown, in this case represented by industrial machinery and locomotion: "A otro mundo de fuerzas productoras/Con fe se lanza nuestra reina amada,/Sus carabelas son locomotoras,/Su América, la España inanimada" (Our beloved queen is leaping into a different world of productive forces/her caravels are locomotives,/her America is the inanimate Spain). While one Isabel relied on Columbus's vessels to ensure the future, the other has locomotives; whereas one discovered America, the other rediscovers an "inanimate" Spain that needs the dynamism of modernization. Both the discovery of America and the railroad expansion are presented as landmarks in the history of the country, achievements that would have been impossible without a divine auspice or the monarch's will.

Torres Muñoz de Luna's poem embodies a gesture of appreciation to the queen for her role as facilitator and promoter of modernization. His panegyric, in that regard, differs greatly from King Juan Carlos's speech, in which it is the monarch who extends his appreciation to the people for their hard and essential work in making industrial progress possible. This symbolic inversion of the relationship between the monarchy and the people suggests a profound transformation of society and, consequently, of national identity—palpable evidence of a long but effectual process of ideological negotiations and technological development that began materializing during the nineteenth century.

In addressing the problems of nation and national identity on the peninsula, recent historiography treats the early nineteenth century as a turning point in the consolidation of Spain as a sociopolitical and cultural unity. While religious, geographical, and ethnic elements had become part of a protonationalism as early as the late fifteenth century, these studies consider that the sentiment of belonging to a nation (understanding the word in its modern meaning) did not consolidate until the Napoleonic invasion in 1808.[10] It was at the outset of the French occupation that Spaniards became aware of shared aspirations, common traditions, and collective beliefs that opposed those of the occupiers. Underwriting the country's institutional transition toward a constitutional monarchy, the proclamation of the 1812 Cadiz Constitution strengthened national sovereignty amid the international conflict with France. One of the main achievements of this document was the re-evaluation of the core political conceptions on which the model of national integration had been built. At the same time, the Constitution upheld certain social structures and cultural practices identified as part of the national essence, which further complicated the articulation of scientific, technological, and industrial modernization. Tensions between traditionalist and progressivist views of the country in this context divided society and questioned the extent to which the national character needed to be defined in relation to allegiance to the monarchy and the profession of Catholicism. This conflict contributed to the political antagonism, social instability, and economic frailty that would later characterize Fernando VII's and Isabel II's reigns during the first portion of the century.

The 1868 liberal revolution changed the balance of these ideological conflicts, yielding the adoption of a slightly more progressivist agenda. Yet the revolutionaries were unable to revamp the country's administrative apparatus—the constitutional and parliamentary system they proposed encountered strong opposition from conservative sectors who saw in these attempts at renovating the country a dangerous threat to their own socioeconomic standing. As a result, as historian José Álvarez Junco explains, liberals found themselves with the dilemma of having to redefine the "Spain" that was best suited to their political project. At the very least, they had to be sure this notion was not delineated exclusively by inherited religion, loyalty to the king, or the traditional values of the aristocracy;

without questioning its unity and power, they needed this idea of the country to function as the basis for constructing a modern state and a participative government.[11] Only six years after the revolution, the system that liberals had put in place collapsed, giving way to the restoration of monarchy and the instauration of an accorded alternation in power of conservatives and liberals. Subject to a political model in which opposing ideologies could in principle coexist, from 1875 onward Spain lived through a period of institutional stagnation that progressively eroded the opportunities for national modernization.

As the ideologist of the Restoration's political structure, historian and politician Antonio Cánovas del Castillo embodied the ideals of a rather conservative liberalism that was reluctant to favor the country's social and economic reorganization. An objector to principles such as free trade and universal suffrage, in his plans Cánovas undervalued the potential of industry's social changes to impact the national economy. In fact, his political agenda operated at the margins of increasingly problematic phenomena, such as social mobility, urban expansion, and population displacement. From Cánovas's perspective, as historian Juan Pablo Fusi explains, there was no doubt that Spain constituted an integrated entity on the basis of its history, geography, and language.[12] Connections between identity, religion, and traditional forms of social stratification were therefore common to his vision of the nation. Yet Cánovas's projects of modernization included attempts to reconcile conservative and liberal convictions. By 1897, however, when he was assassinated, presumably by his political enemies, his system of concerted alternation of parties as heads of the government had already exhausted all its possibilities for harmonizing conflicting ideologies.

After Cánovas's death, less conciliatory political visions emerged that in any case had to focus on alleviating growing social tensions. The new administrations worked on eliminating the government's excessive centralization and on combating the institutional vices that the mechanism of political alternation generated. With these goals in mind, in 1899 Francisco Silvela, representative of the most conservative faction of Canovism, assumed the presidency of the country; and a few years later, in 1902, Alfonso XIII was crowned king. Two visible representatives of conservative ideals were thus in command of a country that needed an injection of modernity if it were to leave its backwardness behind and recover from the moral, economic, and social debacles of defeat in the 1898 war with the United States. This loss, along with the growing unrest of the working class, radicalized the country's ideological division even more, thus preventing a much needed national consensus to promote, among others, the educational and institutional reforms that were required to bolster national modernization.

This pessimistic atmosphere contrasted with the hopefulness that the county's new leadership and the beginning of a new century aroused in certain sectors of society. In a special issue of the journal *La Energía Eléctrica* (Electric energy) published in 1902 and entitled *La ciencia y la industria eléctrica en España al subir al trono S.M. el Rey Don Alfonso XIII* (The science and industry of electricity in Spain at the advent of H.M. the King Alfonso XIII), for example, military engineer, science popularizer, and educator Francisco del Río Joan celebrated the advent of the new monarch as a symbol of regeneration and change:

> Parece como si al cerrar la diez y nueve centuria, se abriese para España una era de consoladora rehabilitación. Parece como si al tocar en un mínimo de la curva social, reuniéramos nuestras fuerzas para ir ascendiendo á un máximo de velocidad. Parece que al advenir un Rey, le acompaña la aurora, y el horizonte se arrebola y se dilata. Parece que la Infinita Misericordia se apiada de nosotros, y que una voz apocalíptica nos grita como á otro Lázaro: —¡Levántate y anda![13]

> (It seems as though the closing of the nineteenth century ushered in a new era of reconciling rehabilitation for Spain. It seems as if at the moment we reach a minimum on the social curve, we gather our energies to start ascending at maximum speed. It seems that the advent of the King was followed by dawn, and the horizon becomes redder and wider. It seems that the Infinite Compassion takes pity on us, and that an apocalyptic voice commands us, as if we were a new Lazarus: —Come out!)

A closer examination of this passage offers clues about the contradictory nature of this period. The optimistic tone used by del Río Joan can be explained in part by his own affiliation with regenerationism—an intellectual and political movement that sought to rescue Spain from its social, economic, and moral decline. Indeed, the idea that the country needed to hit rock bottom in order to start ascending again was one of the arguments put forward by this group of thinkers.[14] Del Río Joan's faith in the young monarch's ability to bring stability and progress to Spain, for example, is represented by the metaphor of dawn. By highlighting that Alfonso XIII's arrival was a breakpoint in the decay of the country, the author thus projects both a realistic assessment of the present and an optimistic view of the future. Through divine intervention (the "Infinite Compassion" suggested in the passage), Spain could be resurrected, a gesture that would confirm the prominent role of religion and the monarchy (symbols of the past) as engines of the future. Such contradiction, as this book shows, is symptomatic of the multiple ideological hurdles that marked national modernization.

Looking at the Past

Before I lay out the various parts that this study comprises, a brief example may be useful to illustrate how the multiple tensions upon which national identity was negotiated during this period—materialism and spirituality, past and present, and tradition and progress—complicated the assimilation of the industrial phenomenon into the country's imagination. Spain's eclectic intellectual environment at the turn of the twentieth century saw the convergence of multiple, and sometimes irreconcilable, philosophical currents—positivism, Krausism, Darwinism, and socialism, among others ideological frameworks, supported attempts at understanding the complexities of progress.[15] Subsumed under the possibilities offered by this broad theoretical spectrum, in a talk delivered at the opening of classes at the Ateneo de Madrid (Madrid Athenaeum) in 1882, Cánovas criticized the effort and energy society devoted to censure the materialist facets of scientific inquiry. For him, this philosophical dispute, along with recriminations of science for placing reason over religious dogmas, had diverted thinkers' attention from more urgent and pressing problems. Cánovas's unease was directed at the unanticipated enthusiasm that Auguste Comte's positivism or Arthur Schopenhauer's pessimism had generated in Spain. Both platforms had destabilized scientific and philosophical endeavors, secularizing their approach to fundamental questions such as the constitution of matter, the nature of movement, the origin of language and reasoning, or the essence of free will. Cánovas concluded that, as long as analytical efforts concentrated on separating the spiritual and material aspects of reality, neither philosophers nor scientists would ever be able to find an answer to these problems. In his discussion, for instance, he referenced neo-Kantian philosopher Charles Renouvier to emphasize that, even when physical laws were accepted as valid, the universe could not be reduced to a simple mechanism. In synthesis, science could only be useful insofar as its philosophical substrate encompassed both materialism and spirituality, an analogous conclusion to that postulated years later by Miguel de Unamuno regarding the rational grounds of faith.[16]

Beyond proposing a basis for the public debate on the role of science in society, the tension between materialism and spirituality also led to fruitless attempts at rendering compatible the most recent scientific discoveries with the Catholic doctrine. As a matter of fact, the cult of nature that spread across Europe, and the conceptualization of science as the mechanism to unveil its secret laws,[17] encountered strong opposition in Spain precisely because of the ideological conflict between the foundations of scientific inquiry and the principles of Catholicism. As historian of science Agusti Nieto-Galán has noted, scientific views of society popular amid the progressive convictions that fueled the 1868 liberal revolution were later rejected and even proscribed, once the conservative guidelines behind the Restoration called into question the validity of recent scientific postulations.[18] French positivism and German Naturphilosophie illustrate two different approaches to this tension and serve as antecedents in understanding the complexities of the Spanish case. In Naturphilosophie, physical laws had to be derived from nature, with science being understood as a systematic procedure for exploring the creations of God. In positivism, physical laws prevailed upon nature, and science was seen as the set of procedures for discovering and making sense of those principles, their properties, and their relations. Both approaches oscillate between subjectivity and objectivity, between metaphysics and materiality. The emphasis of German science on the organic greatly differed from the stress the French system gave to the causal and mechanical character of nature. Analogously, cultural appropriations of technology and industry in Spain fluctuated between these two contrasting forms of scientific interpretation.

Significantly, the debate around scientific materialism also had repercussions on the organization of higher education. In Spain the Catholic Church had authority to decide on the content, teaching methodologies, and lines of study available in universities. It is not a coincidence that during the last decades of the century, educators' and scientists' inaugural talks as members of the Real Academia de Ciencias Exactas, Físicas y Naturales (Royal Academy of Natural, Physical, and Exact Sciences) insistently focused on the state of science and the way freedom of thought was affecting education. The popularization of scientific knowledge also raised awareness of the adverse repercussions that a close relationship between the Church and state could have on education. Part of the problem was the adherence of Spanish identity to Catholicism, which had prevented a questioning of tradition indispensable to consolidating the country's projects of modernization.

Not surprisingly, in this context many ideas proposed by materialist scientists and thinkers, including Ludwig Büchner, Charles Darwin, and Herbert Spencer, became serious threats to the system of values that supported the national ethos. Thus, while positivism, evolution, and anthropocentrism were linked to negative effects of modernization, Naturphilosophie got to be more appealing, as it facilitated the reception and adoption of less dogmatic philosophical edifices, such as the one advanced by German philosopher Karl Christian Friedrich Krause. Imported to Spain by Julián Sanz del Río around 1850 and widely disseminated among a group of liberal intellectuals during the revolutionary period, this doctrine became pivotal in their political efforts to execute social and institutional reforms. As Juan López-Morillas explains, in Spain Krause's ideas transcended the theoretical and philosophical debate to become part of a political and even religious discussion: "Francamente confesamos que, más que el análisis de un sistema filosófico, nos atrae la caracterización de una modalidad cultural. Lo que en el Krausismo quiso ver cierto tipo de español se nos antoja mucho más significativo que lo que Krause puso en su doctrina o que lo que Sanz del Río se propuso al importarla" (In all honesty, I have to say that more than an analysis of a philosophical system, my study is a characterization of a cultural mode. What certain Spanish thinkers wanted to see in Krausism is much more significant than what Krause proposes in his doctrine, or what Sanz del Río expected when he imported it).[19] As an intellectual platform, Krausism offered the perfect model for shunning three different views of Spanish identity. First, the idea that Spaniards were unable to produce or understand abstract notions and, as a consequence, were incapable of generating scientific or philosophical knowledge—the national genius was then inclined to an imaginative exuberance that highlighted their extreme individualism and passionate spirit. Second, the notion that this particular character was a response to the religious radicalization of society that had taken place after the Counterreformation. And, finally, the perception that in the aftermath of this schism the Spanish personality was consecrated to Catholicism, entrusting the Church with the mission to restore the country's social and moral orders. One consequence of these perspectives was the consolidation of a Manichean approach to religion: any ideological tendency that threatened the Church's sovereignty or seemed heterodox was considered anti-Spanish.[20] Krausist adherents, for their part, developed an original vision of the country built on the objective revision of reality—one possible path toward modernization consisted precisely in the rationalization of culture. In this light, Spain seemed to be suffering from a *disease* for which the lack of scientific, technological, and industrial development was one of its most severe symptoms.

The emergence of a rhetoric of illness was indeed part of a larger project of rationalization of the national problems. The function of the discourse of degeneration, as Michael Aronna explains, "was to determine which groups and practices constituted biological and cultural obstacles to modernity, to diagnose the illness afflicting these groups and to develop treatments or solutions."[21] In assimilating technological and industrial change, concepts such as decay, degeneration, and disorder also functioned as markers for assessing social practices, traditions, and identity traits that could represent serious obstacles to the projects of national modernization. As objective and effective tools with which it was possible to *diagnose the nation*, analysis and rational interpretation thus reaffirmed the importance of education for social advancement.

Within Krausism, education was considered indispensable for putting national problems into perspective. As a matter of fact, Sanz del Río had a rather specific idea of the role universities played in the country's process of modernization:

> No se debe pensar—escribía [Sanz del Río] desde Heidelberg en 1844—que universidad es y significa en Alemania lo que en España. Nuestras universidades son instituciones donde se enseña la ciencia, antiguamente bajo la influencia y aun dirección eficaz, directa, íntima de la Iglesia, y ahora del Estado; en Alemania la universidad es en su interior, en la enseñanza misma, una institución totalmente independiente de la Iglesia y del Estado; con tal que sea verdaderamente ciencia lo que en ella se enseña.[22]

> (We should not assume—he wrote from Heidelberg in 1844—that university is, or even means the same in Germany as in Spain. Our universities are

institutions where, in the past, science was taught under the Church's intimate, direct, and effective supervision, and in the present, under the state's vigilance. In Germany, regarding education and instruction, the university is a completely independent institution; at least with respect to science.)

Leading Krausists, such as Francisco Giner de los Ríos, Nicolás Salmerón, or Gumersindo de Azcárate, considered that it was time for universities in Spain to seek support in the society they were supposed to serve, and not in the state that had kept them in a condition of serfdom. Azcárate, in particular, thought that university independence and vitality were closely linked, and argued that any institution subjugated to an external administration would be unable to make effective decisions. In a similar fashion, Salmerón contended that the only regulation that should apply to the university operation should be one that guaranteed free inquiry and constant search for the truth—since this was a social goal, educational institutions then needed to be accountable to society. In practice, however, intellectual flourishing was jeopardized by the state's constant attacks on academic freedom.

The creation, in 1876, of the Institución Libre de Enseñanza (Independent Institute of Education) responded precisely to criticism coming from the government and the Church. Specifically, it was a reaction to the royal decree promulgated by secretary of education Manuel Orovio restricting instruction that could contravene Catholicism. The institute, as historian of science Stefan Pohl-Valero explains, was aimed at educating "a liberal elite destined to reform and modernize Spain."[23] Founders and educators affiliated with this project thus endorsed progressivist and democratic principles akin to Krausism, such as ideological independence and secularization.[24] In fact, Krausist philosophy inspired the Institución's teachers, who strived to consolidate a group of leaders capable of addressing the country's most urgent problems.[25] Within the scope of this mission, writing was deemed a crucial activity, one that not only ensured freedom of inquiry but also warranted an objective, almost scientific diagnosis of society. One of the greatest advantages of the Krausist approach to education was its balanced perspective on spirituality, in which both individual responsibility and religious values were taken into account. As Roberta Johnson explains, Krausism laid the metaphysical and practical groundwork upon which debates on the relevance of scientific research for the country's progress and the development of an authentic scientific tradition consolidated.[26] This was particularly relevant for Spanish thinkers who embraced the materialistic aspects of modernization but defended Catholicism as one of the central components of their identity.

Framed within this context, Cánovas's talk at the Ateneo not only encapsulates many of the ideological tensions that complicated the shaping of industrial imagination in fin-de-siglo Spain, but also suggests the heterogeneity and complexity of scientific and technological assimilation, highlighting the interconnection among different forms of knowledge, spaces of discursive production, and ideologies. As he emphasizes in the conclusion, his goal was to present a scenario in which it was possible to integrate the "new science"—that is, the study of society in light of technological and scientific advances—and the notion of a nation inscribed in the universal order of Catholicism:

> Quise en breves términos describir el estado de la filosofía contemporánea, y confesaros al principio mi esperanza íntima de que *la nueva ciencia*, que por fuerza ha de reemplazar algún día esta atonía filosófica en que al presente estamos, deje de una vez aparte, como hecho manifiesto é indestructible, la existencia de un *orden universal*, subjetivamente inteligente, previsor, omnisciente.[27]
>
> (In brief, I wanted to describe the state of contemporary philosophy, and I confess, from the beginning, my intimate hope that the *new science*, which necessarily must some day replace the philosophical dissonance of the present, once and for all reveals, as palpable and indestructible, the existence of a *universal order*, subjectively intelligent, foresighted, omniscient.)

Unsurprisingly, Cánovas was not the only author whose ideas on the nation, national identity, or society in general responded to the new epistemological and cognitive parameters imposed by science, technology, and industry. In exploring some of these cases, *Founders of the Future* demonstrates how modernization's symbolic and material resonances transformed the entire space of cultural production.

A Different Perspective

In its evaluation of a broad archive of cultural practices, this book joins numerous theoretical, historical, and critical discussions about the social and cultural impact of industrial modernization in general, and about the textual negotiations of national identity in fin-de-siglo Spain in particular. Material cultural studies, for example, have tackled the problem of representation that surfaces at the interface between society and objects. Authors such as Nicole Boivin or Ismael Sarmiento, among others, have shown how an object can have multiple connotations in different societies and, as a consequence, how different narratives emerge in response to these particular experiences of materiality.[28] Likewise, material culture critics, such as Ian Woodward, have specifically analyzed the connections between technology and society to show how they problematize the conceptualization of national identity.[29] Following these ideas, *Founders of the Future* suggests that industrial development generated its own allegorical repertoires for addressing society's conflictive response to the reformulation of tradition.

The historiography of industrialization in Spain has mainly focused on analyzing the political and ideological tensions produced by the economic and social transformation of the country during the nineteenth century. Authors such as Jordi Maluquer, for instance, have shown how Spain's late and unstable industrial development responded to inefficient state administrations that, in his view, implemented diffident or misconceived policies at the expense of social and economic improvement.[30] Similarly, historians such as Antonio Gómez Mendoza or Gabriel Tortella associate these problems with a lack of adequate regulations for the promotion of science and education, the protection of local industries, or the stimulation of investment.[31] In general terms, these approaches have looked closely at the role played by different political and social actors in the consolidation of national projects that dealt with the new challenges of industry. The emerging bourgeoisie, for example, has been at the center of many cultural analyses that explore the transformative capacity of material modernization. As I show in the first chapter, cultural critics and historians have studied the interrelationship between literature and social change in the context of mass production and other phenomena derived from the accelerated and uneven development of the country during the second part of the nineteenth century. Regarding the rapid growth of European economies as one of the causes of the disarticulation of traditional social structures, cultural critics, including, among others, Stephanie Sieburth, Jo Labanyi, Francisco Caudet, and Jesús Cruz, have pointed to the reconfiguration of the public and private spheres, and the emergence of new strategies of social control that responded to the exceptional conditions of modernization in Spain.[32] From these perspectives, discursive production is seen as a figurative vehicle that channeled authors' fears, expectations, and ideas about the country.

Notwithstanding their important contribution to the study of late nineteenth-century Spain, these groundbreaking analyses overlook the profound semiotic changes elicited by modernization. Labanyi, for example, relates the consolidation of the Spanish realist novel to contemporary debates in economics, medicine, and town planning, among others, but does not delve into the emergence of a new cognitive structure that changed how authors saw national modernization. Sieburth, for her part, focuses on the contradictory nature of Spanish modernization—a process in which expressions of modernity such as consumerism and mass culture coexisted with older ways of life. Yet she does not consider industrialization as one of the encompassing symbolic forces that contributed to perceptions of unevenness in the country's process of transformation. Industry and its semiotic repercussions are not the focus of Cruz's or Caudet's studies either. Cruz's enlightening analysis traces the development of a particular lifestyle ascribed to the middle classes and shows how it contributed to Spanish modernity; however, it does not address industrial change directly as one of the symbolic and material forces that facilitated the configuration of such ways of living. Finally, Caudet is interested in the development of literary movements, such as naturalism or realism, as cultural responses to the social reorganization of modernization. As a consequence, his work focuses on the literary forms and styles that evolved as discursive mechanisms to negotiate the social and political changes that industry's economic prosperity triggered in Spain but takes no notice of the cognitive aspects involved in this process.

All these contributions to the study of modernization in Spain coincide in their view of industrialization as a material and ideological phenomenon that complicated the negotiations of national identity. Within this context, discursive production is approached as a reflection on the effects that scientific and technological development had on society. Yet industrialization had deeper ramifications—my claim that material modernization led to a paradigm shift in the representation of the national problems thus seeks to transcend readings of cultural production that see a direct relationship of causality between the evolution of science, technology, and industry, and the emergence of certain topics, narrative structures, and ideological conflicts during this period. Actually, as stated earlier, *Founders of the Future*'s main goal is not to provide a historical account of the process of industrialization in Spain, or a critical review of its impact on cultural production in general, or on political, economic, and social discourses in particular, a task other critics have undertaken successfully.[33] Although the present book draws extensively on that literature, its aim is to show how cognitive and material aspects of industrial progress shaped the discursive mechanisms and figurative strategies that authors deployed in their attempts at coming to grips with the country's transformation.

Founders of the Future's overall structure reflects this interest in examining multiple forms of discourse to present a larger and broader picture of the institutional challenges and cultural struggles that Spain confronted during its complex process of industrial modernization. The book's various chapters reconsider concepts such as nation, progress, and development within the semiotic substructures made available by processes of economic expansion, educational improvement, social reform, circulation of knowledge, and environmental change. Chapter 1, "The Social Foundry," presents a conceptual framework for reading and interpreting the technological metaphors, scientific allegories, and industrial images that fin-de-siglo Spanish thinkers used in their evaluations of the country. Presenting the steel industry as a referent in the analysis of society, the chapter delineates the concept of social foundry—its allegorical nature, material character, perceptual component, scientific dimension, and historical condition. An exploration of the symbolic structures in which the collective experiences of the industrial phenomenon yielded new forms of imagining and representing reality reveal the fluid dialogue between industry and cultural production during this period. This section also provides a brief historical overview of the main problems associated with the conflictive process of industrialization in

Spain, emphasizing aspects such as economic variation, social conflict, and environmental and human cost.

Chapter 2, "Economy and Other Matters of State," explores interconnections among physics, geology, and economics during the last decades of the nineteenth century. Focusing on three authors—scientist and politician José Echegaray, geologist Lucas Mallada, and economist Pere Estasen—the chapter shows how their analyses of the national financial system were built on scientific notions of industrial productivity. Within this context, ideas of energy conservation, power dissipation, or resource depletion became common referents in their understanding of the forces that regulated the economy. Their social foundries therefore consisted in an assessment of the national wealth that highlighted the importance of science and education for strengthening society's multiple attributes. From this perspective, it was essential for state administrators to know and understand the country's physical dimensions and the quality of its natural and human resources in order to properly articulate a defense or critique of economic platforms, such as protectionism of free trade. In their roles as scientists, educators, and writers, the authors discussed in this chapter created links among science, industrial development, and political economy, showing the interpretative value of physics, geology, and economics as a referent for the diagnosis of the ailing national body.

Chapter 3, "The Educational Engine," reviews two different areas. On the one hand, it deals with scientific concepts, their reception, and their assimilation in Spain. On the other hand, it analyzes how these discoveries provoked a re-evaluation of practices in scientific instruction. Technical development was associated with economic modernization and social progress. Due to historical and ideological reasons, however, the country had encountered many obstacles in its attempts to develop an original science that contributed actively to the consolidation of industry. For many scientists, educators, and politicians, the system of higher education and the multiple deficiencies in its organization and administration were to blame for this setback. Without an adequate educational structure, by means of which practical and theoretical instruction were tuned to the country's needs, the consolidation of a Spanish scientific tradition was unlikely. I analyze how proposals for curricular and structural reform also became complex spaces of ideological negotiation that were instrumental in the attempts to reconcile materialism and spirituality. The chapter focuses on essays and public interventions by educators and scientists Laureano Calderón, José Rodríguez Mourelo, Gurmersindo Vicuña, and Zoel García de Galdeano. It shows how industrial imagery became central to their proposals and perspectives on higher education.

Chapter 4, "Social Engineering," examines the effects of industrial transformation on labor practices, the consolidation of the proletariat as a social power, and the displacement and relocation of workers in fin-de-siglo Spain. I explore how this impact was symbolically negotiated in literature through the depiction of modernization as a socially disruptive force. With the motif of technological misappropriation as a background, the novels studied in this chapter criticize the use of nationalism or religion as part of ideological agendas that sought to protect power structures, preserve social differences, and prevent a working-class uprising. This criticism adopted the form of a social foundry in which industrial processes were used as referents to justify or condemn social disparity, labor exploitation, and capital accumulation. Through the analysis of works by writers Vicente Blasco Ibáñez, Manuel Ciges Aparicio, and Concha Espina, the chapter explores the new forms of internal colonialism that a profit-based economy instigated in the country.

Chapter 5, "Technologies of Mass Diffusion," focuses on the dissemination of scientific ideas, the popularization of technology, and the role of these cultural practices in the evaluation of society. Many writers and journalists of the period brought into public debates some of the most important scientific theories being developed in Europe and successfully deployed in the industrial realm. Their main interest was to consolidate a space to confront materialism and spirituality, and to defend tradition against modernization. Most of these interlocutors felt that it was necessary to reject discoveries that threatened religious beliefs in spite of their material benefits for the country. Others thought it essential to spread these ideas and appealed to the state for the modernization of education and the promotion of science. The chapter explores examples of each of these kinds of popularization in works by Emilia Pardo Bazán, José Echegaray, and Pedro Antonio de Alarcón. Discussing topics such as thermodynamics, evolutionary biology, or transportation, these texts articulate and discuss their own visions of the country, social foundries that in many cases helped audiences to think about the role of science, education, and technology in constructing the modern nation.

Chapter 6, "Industrial Footprint," considers the environmental and biological effects of industrialization. I show how writers used fiction to incorporate new strategies of representation that reflected the conflictive appropriation and problematic consequences of material modernization in rural Spain. In the countryside, the institution of industrial regimes dislocated traditional social structures and subverted power dynamics. Industrial expansion also had a profound impact on nature by defacing the landscape and depleting the soil. In rural areas, modernization was thus intrinsically associated with the symbolic destruction of the past. In this chapter, the analysis of works by writers Benito Pérez Galdós, José Ortega Munilla, and Armando Palacio Valdés provides yet another perspective on the unstable and contradictory push toward modernity in Spain.

The different analyses presented in this book offer the possibility of reevaluating industrial modernization as a semiological engine that produced new images of the national ethos. By studying the particular assessments of society generated at the interface between industry and ideology, *Founders of the Future* proposes an amplified canon of fin-de-siglo Spanish cultural production whereby to explore the country's complicated and in some ways detrimental negotiation of its identity.

CHAPTER 1

The Social Foundry

Scientific, technological, and industrial development had a far-reaching impact on Spanish society. Beyond their political and economic implications, these transformations also engendered new forms of symbolic representation. Take as an example an event that left a profound impression on the writers and journalists of the time: the arrival, on February 9, 1851, of the first train at Atocha Station in Madrid.[1] Many articles about the occurrence were published days later on the front pages of newspapers such as *La España* (The nation) and *El Heraldo* (The herald), where depictions of the moment enthusiastically celebrated the triumph of technology—the train's arrival had become a metaphor for the dawn of progress in the country, as Ángel Fernández de los Ríos would highlight later in his 1876 *Guía de Madrid* (Guide to Madrid):

> Un día, el 9 de febrero de 1851, al lado del convento de Atocha apareció un monstruo que vomitaba humo, sembraba fuego, bramaba cien veces más fuerte que el león del Retiro, hacía llegar un silbido a medio Madrid, arrastraba cincuenta carruajes en que cabía la carga de todos los simones de Madrid juntos, y devoraba el espacio más que todos los tiros de mulas de Fernando VII desbocados.[2]
>
> (One day, February 9, 1851, a monster appeared alongside the convent at Atocha that vomited smoke, spewed fire, and roared a hundred times louder than the lion at Retiro; its whistle reached half of Madrid, the fifty cars it dragged behind it could carry the cargo of all the carts in Madrid put together, and it devoured more space than all of Fernando VII's mule teams at full gallop.)

The repertoire of images incorporated by Fernández de los Ríos in this passage merits closer examination, as it shows how the various physical manifestations of modernization affected and transformed the writer's poetic imagery.

Fernández de los Ríos's description simultaneously reveals the transformation of society and the reconfiguration of the narrative strategies employed to imagine it. The visual, auditory, and spatial-temporal allusions in the text not only indicate a disruption of perception caused by technology but also point toward the political, social, and economic problems Spain faced at the time. The image of the machine as something monstrous, for example, demonstrates the extent to which there remained an incompatibility between the cognitive principles used to represent the past and the referential materiality of the present. In Fernández de los Ríos's invocation of the *monsters* of modernization, rationality prevails over primitive ignorance and superstition, and technology acquires an almost metaphysical role in society. By describing the sounds that the train produced as animalistic and intimidating, the journalist was simultaneously juxtaposing the capacity of nature (the actual lion at the Retiro park zoo) with the reaches of man-made objects and surreptitiously alluding to the monarchy ("the lion at Retiro"), indicating that modernization was unstoppable and much stronger than Fernando VII's conservative ideology and the obstacles it had posed for the country's development. Indeed, the comparison made in terms of velocity between the temporality of traditional transportation ("the mule teams of Fernando VII") and that achieved by the train—a clear image of technological progress—provides a striking example of the semiological impact of modernization.

Taking as a point of departure the reconfiguration of discourse that came about during the technological and scientific revolutions of the nineteenth century, this chapter presents a critical framework to approach industrialization as a space of cognitive transformation that gave shape to a new way of approaching reality—the complex system of representation that I have termed *social foundry*. I review its semiotic and allegorical nature, its material character, its perceptual component, its scientific dimension, and its historical condition. A social foundry operates under the same principles that govern the appropriation of scientific notions in fields like economics, politics, or literature. These "scientisms," as cultural critic Bruce Clarke defines them, "generated in their turn vehement, often visionary counternarratives and critical responses, and these significantly shaped the creative literature and imagery of the period."[3] In their analyses of the intersection of science and culture, historians and cultural critics of industrial modernization have particularly concentrated on this rich semiotic and semantic exchange.

Social foundry's articulating principles, in fact, draw on the ideas of cultural critics, such as Laura Otis, Phillip Mirowski, Nicholas Daly, Ted Underwood, Barri Gold, and Tamara Ketabgian, whose work offers an original insight into the metaphorical uses of science and technology for describing and evaluating reality.[4] Otis, for example, is concerned with the ability of scientific discourse to characterize society and of common language to explain science, a preoccupation that Mirowski also evaluates in his discussion of the role of physics in the development of modern economics. In his view, the emergence of liberalism and the debates on protectionism and free trade in Europe can be understood in terms of the principles of energy conservation and entropy. For his part, Daly translates many of these concerns to the cultural sphere by studying the way in which representations of technology generated specific subjectivities and helped define modernity in literature and film. Literary production is also the subject of Ted Underwood's study, in which he traces the uses of images related to natural forces, particularly those associated with solar energy, to show how they marked the emergence of discourses on labor and productivity that were common in the nineteenth century.

The emphasis of these works on the participation of different social actors in the process of technological transformation is also key to understanding the complex connections between society and industry. Ketabgian and Gold, for example, look at the industrial and scientific readings of modernization found in Victorian literature, social treatises, and scientific texts, and show how these discussions proposed specific ways to address the numerous social problems of the period. While Ketabgian investigates the realm of emotions, looking at the feelings of alienation and reification that the mechanization of industry inflicted on workers, Gold focuses on thermodynamics and its ubiquity within the Victorian cultural imagination. Although these studies inform my discussion of the epistemological and semiotic implications of the nineteenth-century proliferation of industrial referents, my conceptualization of the social foundry also draws attention to ideas of human agency and transformative capacity, thus expanding the scope of these approaches to the encounter of science, technology, and culture.

Multiple scholars have addressed the impact that science and technology had on Spain's projects of national modernization. Historians of science José Manuel Sánchez Ron and Stefan Pohl-Valero, for example, have reviewed the connections between science, politics, and culture in the country.[5] Paying particular attention to the political appropriations of physics and its repercussions on the social and cultural imagination, Pohl-Valero has advanced the idea of social thermodynamics as one possible frame of reference used by authors of the period to analyze the country's various social and economic problems.[6] Along with thermodynamics, the popularization of evolutionary biology and the adoption of Darwin's principles as models of social behavior provide important clues about the process of symbolic transformation studied in this book.[7] In fact, these active and fruitful exchanges between science, technology, and society serve as models to relocate industry at the core of the prescriptive endeavor of many writers, educators, scientists, and politicians of the time.[8]

As is shown in what follows, the transformative capacity of industry, condensed in the image of the foundry, became a powerful rhetorical device to address problems related to social, economic, and institutional progress. The mechanisms involved in the industrial process of forging metals not only synthesize late nineteenth-century industrialization as a whole (mining, railroad expansion, and so forth), but also feature physical concepts, such as energy and work, as operative referents in the evaluation of society. One advantage of the popularization of science during this period, as Greg Myers explains, was the assimilation of complex ideas, concepts, and theories into quotidian experiences, generalizations, and commonplaces; in doing so, writers were able to adopt scientific developments and technical advances as referential repositories for their abstractions of reality.[9] This exercise is what gave rise to models like social Darwinism or social thermodynamics. The same can be said of industry; yet the social foundry as a framework for describing social transformation in industrial terms takes this logic one step further by considering the way in which cultural and social agents not only appropriated scientific notions and technical elements to interpret the country's problems but also used these organizational principles to make sense of identity's malleable nature.

The Material Field

To better understand the relationship between industrial modernization and the emergence of figurative mechanisms to make sense of social transformation, two conceptual domains need to be considered: that of material culture, put forward by authors such as Rom Harré or Christopher Tilley, and that of language and representation, as advanced by linguists George Lakoff and Mark Johnson. The main advantage of these theoretical paradigms is their understanding of discursive production as the outcome of historical, social, and economic factors, as well as the result of material conditions associated with modernization. Metaphor, as a fundamental figure of speech, is a determining element in the cognitive configuration of new systems of signification. As Lakoff and Johnson explain, metaphor is much more than a poetic or imaginative literary device: "New metaphors, like conventional metaphors, can have the power to define reality. They do this through a coherent network of entailments that highlight some features of reality and hide others."[10] Through the transference of meaning from different symbolic spaces, metaphors allow us to make sense of unfamiliar aspects of the world; a function that, as Tilley observes, is implied in the word's etymology: "*metaphora*: carrying over."[11] This principle of disparity between language and reality also appears in the context of industrialization and can be evaluated in the correlation that exists between the material transformations of industry and their impact on the social, economic, and cultural realms.

If, as Tilley proposes, metaphors provide a mechanism to link "empirically (factually) disparate and unconnected" frames of reference,[12] it is not surprising that this trope became the privileged space for the appropriation of scientific, technical, and industrial concepts in *fin-de-siglo* social discourses. Cultural critics have questioned to what extent this pervasion of images contributed to the understanding of society. Otis, for example, has concluded that in the nineteenth century metaphors were part of a complex network of knowledge that provided necessary referents to address social and cultural tensions.[13] By including everyday interactions, this cognitive transposition became very productive in the categorization and organization of reality.[14] In fact, when analyzing the uses of energy as a mechanism of representation in the era of classical thermodynamics, Clarke points to how "reification, personification, typological allusion, and emblematic juxtaposition form a significant part of the arsenal of tropes and schemes that both shape up and ship out scientific ideas."[15] Similarly, a social foundry consolidates from images generated at the conceptual and functional levels of industry. The physics of energy, and particularly thermodynamics, cannot be overlooked at this juncture, as it created models to explain and even justify different aspects of society within the context of modernization. Preoccupations with productivity and efficiency, for instance, could be better dissected through images that describe energy consumption and mechanization as variables in an integrated system. One of these metaphors was the human motor, a trope intellectual historian Anson Rabinbach links to appropriations of the laws of thermodynamics and their application to the worker's physiology, now seen as one more component of the industrial apparatus. This was an association with "worldly consequences far beyond [its] status as literary expression," a way of connecting notions such as force, fatigue, and exhaustion with the social reality of labor exploitation and other problems of modernization.[16]

The other aspect of industry that needs special attention in the conceptualization of the social foundry is the material. In "Material Objects in Social Worlds," Harré suggests that objects are entities defined through specific social narratives. According to him, these narratives and the construction of meaning that takes place within them are defined by the role things play on two specific orders—the practical and the expressive:

> Human beings have always, so it seems, lived in a double social order. One component consists of the social arrangements for maintaining life in such and such an environment. This is the practical order and people have their locally proper places in such an order. The other component consists of the social arrangements for creating hierarchies of honour and status. This is the expressive order. Material things can be understood in their full human significance only if their roles in both these orders are identified.[17]

To illustrate this double dimensionality, one can take as an example the steel furnace, an artifact central to Spanish industrialization and to the social foundry's imagery. On a practical level, the smelter is fundamentally a tool that efficiently converts raw materials into steel. On an expressive level, the furnace constitutes an emblem of modernization and technological advancement—a token of the human ability to understand and manipulate nature. This association of the industrial production of metals with a society in constant transformation is not arbitrary. At the end of the nineteenth century this became clear, as the operation of the steel industry put into circulation ideas about dynamism and collectivity that shaped the understanding of national identity.

The value of this kind of figurative associations, as Clarke confirms in the case of thermodynamics, lies in its adaptability: "The abstract analogical foundation of this interplay between allegory and energy physics is the shared quality of *transformation*, the possibility of semantic as well as material and conceptual metamorphoses."[18] It makes sense that in a century marked by the instability and constant remodeling of the social order, transformative models of society turned into popular and effective ways of describing reality. More importantly, as Rabinbach asserts, the nineteenth-century scientific revolution yielded "the belief that human society and nature [were] linked by the primacy and identity of all productive activity, whether of laborers, of machines, or of natural forces."[19] Given these conditions, and acknowledging the cognitive dimension of industrial transformation, it is critical to define a conceptual framework that encompasses the multiple interactions between materiality and discourse. Such a structure, which I term *material field*, allows linking questions of identity, social progress, and labor conflict, among many others, with the theoretical and practical complexities of industrial development, thus facilitating the articulation of social foundries.

The material field relies on objects' ability to alter perception. A third dimension to Harré's model of material interaction described earlier, this perceptual level underscores how industrial dynamics distorted notions of time, dislocated conceptions of space, and reconfigured social interaction. In his analysis of the logics of representation, Gilles Deleuze observes how identity is central to processes of cognitive appropriation that take place within the practices of modernization. Here, identity is defined in terms of the temporal and spatial categories of divergence and displacement. The introduction of concepts like *difference* and *repetition* responds precisely to an experience of reality that is always conditioned by markers of time and space. Pointing to the unstable and multiple character of modern society, repetition, as Deleuze understands it, is not linked to the reproducibility of a certain phenomenon, but to its recurrence under exceptional circumstances: "Repetition is a necessary and justified conduct only in relation to that which cannot be replaced. Repetition as a conduct and as a point of view concerns non-exchangeable and non-substitutable singularities."[20] For Deleuze, the contrast between that which is perceived as the static nature of daily life and its actual dynamic and transformational essence evinces the profound impact that objects have on perception. A good example of this duality in the context of nineteenth-century modernization can be seen in the social reconfiguration that occurs inside a train carriage. Deleuze explains that one must distinguish between *difference* defined as "something distinguished from something else"—which he calls *conceptual difference*—and the notion of difference as such: "Something which distinguishes itself—and yet that from which it distinguishes itself does not distinguish itself from it."[21] Every trip in a train between two specific points seems to be identical in terms of its route and schedule; however, it is different in that it (re)creates a physical and imagined space for the passengers that exists as an unrepeatable, unreproducible reality until the trip concludes.

Within these parameters, it can be said that objects establish a tripartite relationship with reality: they are in the world, they constitute cognitive images, and they are mutable entities subjected to time. The way in which objects are experienced and assimilated thus depends on how they interact with society, how they are individually and collectively perceived, and how these two situations evolve over time. Difference, repetition, and temporality are then the main vectors in the cognitive appropriation of an object. Here it might be helpful to think of materiality in terms of what theorist Arjun Appadurai has denominated the "social life" of things, a sort of identity or character that determines the relationship objects establish with society.[22] For Appadurai, this connection is continuously shaped by cultural and socioeconomic factors. In the case of industry's materiality, for example, the practical, expressive, and perceptual levels that mark this interaction hinge on problems such as productivity, labor capacity, efficiency, and so forth. In fact, it was by following this kind of logic that different material aspects of the industrial phenomenon became referents for describing reality and thinking about the national problems.[23]

A representation of social interactions and cultural negotiations whose characteristics are governed by industry's materiality, the material field is also contingent on historical factors. Here the notion of *field*, as advanced by Pierre Bourdieu and Michel Foucault, provides useful elements for exploring the social and cultural practices that emerged or consolidated around the industrial changes of modernization. In Bourdieu's cultural field, for example, social dynamics and political views demarcate the limits of discursive production.[24] In Spain such predispositions led to contradictory ideas on the importance and pertinence of scientific and technological development. While different authors looked for the effective consolidation of material modernization, many of them also rejected its social effects. As pointed out earlier, adducing the deterioration of national identity, conservative and traditionalist sectors stigmatized progress as a destabilizing force. Liberals and progressivists, for their part, proposed mechanisms and strategies to improve the country's social conditions and embrace the positive aspects of modernization.

Harré's concept of "social world" is especially fruitful in this context for understanding how these attitudes interacted with the transformative power of industry and generated the conditions that sustain the material field: "A social world is an ephemeral attribute of a flow of symbolic interactions among active people competent in the conventions of a certain cultural milieu. The major mode of symbolic interaction for modern people is discursive, involving the performance of meaningful actions, such as making gestures, moving material stuff around and shaping it, using linguistic forms, and so on."[25] Within the social world objects can be categorized as passive or active. Things are passive if they do not generate new narratives; that is, if no new forms of abstraction emerge in response to their presence. Conversely, they are active if they engender new mechanisms of representation. In both cases, the relation between subjects, objects, and discourse is primarily based on the construction of meaning and dependent on factors such as political ideology, social status, and intellectual standing.

Similar factors are also at the core of Michel Foucault's conceptualization of discursive field. Foucault coins the term *entities* to refer to everything (concepts, objects, social phenomena) that can be defined by discursive mechanisms. The construction of these entities, however, is subjected to specific principles: "The conditions to which the elements of this division (objects, mode of statement, concepts, thematic choices) are subjected we shall call the *rules of formation*. The rules of formation are conditions of existence (but also of coexistence, maintenance, modification, and disappearance) in a given discursive division."[26] Underlying discourse there is a historical substratum that serves as its (a priori) foundation and determines its possibilities. Discourse can then be defined as a group of statements that constitute knowledge at any specific point in history.[27] This does not imply, of course, that the definition of discourse is purely linguistic; on the contrary, discourse, as it is proposed here, alludes both to language and practice, to enunciation and action—it is everything that can be expressed; it, therefore, exists beyond the text, including its conditions of production and the edifice of knowledge that these constitute.

Foucault thus alludes to a *discursive formation* when cognitive referents across different narratives (a book, the complete works of an author, a scientific article, and so forth) share a set of political, social, or, more generally, contextual conditions. Discursive formations depend on the type of enunciation—scientific treatises, for example, differ from historical studies, novels, and so on—but actively intervene in the ongoing discussions of a given discipline: "But we must understand by [discursive formation] the totality of things said, the relations, the regularities, and the transformations that may be observed in them, the domain of which certain figures, certain intersections indicate the unique place of a speaking subject and may be given the name of author."[28] These conceptual considerations are also relevant if one thinks about the negotiations that take place within the material field. The purpose of Foucault's archaeology of knowledge is not to approach discursive formations as mirrors of reality; that is, "it is not an interpretive discipline: it does not seek another, better-hidden discourse. It refuses to be allegorical."[29] Rather, its goal is to recognize the relations that make it possible. Likewise, studying discursive production within the material field of industrialization does not imply a revision of the way in which industry, technology, or science is depicted; it is, instead, the exploration of the various conditions (epistemological, material, historical, and so forth) that made these experiences possible. A cultural analysis of industrialization following this approach is therefore not restricted to an evaluation of modernization; it also comprises a study of the different instances in which its systems of representation were reformulated.[30]

While Bourdieu's emphasis on the position and predispositions of cultural actors reveals the importance of political ideologies and social conditions for the production of discourse, Foucault's archaeology offers an insight into the symbolic mechanisms that facilitated its occurrence. At the confluence of these views, the material field provides an ideal framework for exploring the semiotic and practical dimensions of industry, as it became a constitutive element of culture at the end of the nineteenth century in Spain. To a large extent, the social foundry owes its prescriptive power to the way in which these two facets altered the perception of reality.

Materiality and Perception

By materializing different mechanisms of interaction between science and technology, and between technology and culture, industry builds what Michel Serres denominates *opérateurs spatiaux* (space operators).[31] Working at the semiotic level, these operators form new semantic and conceptual corpuses that can be utilized to reflect on reality. Yet the impact of industrialization is not merely conceptual. The metaphorical conventions that underpinned the discursive spaces of modernization were also the consequence of a reconfiguration of the perceptual world. As proponents of material culture suggest, objects' physical properties have profound effects on the construction of social narratives. In fact, "social" objects, as Harré qualifies them to highlight this connection, are framed within particular symbolic categories. *Flag, dollar, industry,* for example, have specific meanings and implications in different contexts—in other words, there are precise principles (historical, economic, cultural) that condition the sensorial assimilation of material realities.

While the industrial expansion had profound repercussions on social and economic structures, altering all aspects of daily life, at the sensorial level it particularly affected the perception of time and space. In his sociology of the emerging industrial metropolis, Georg Simmel alludes to this phenomenon as the "assault on the senses" of modern life.[32] Consequences of this *assault* are reflected, for example, in the increasing interest in measuring and utilizing time with efficiency and precision.[33] Detailed train schedules and efficient lines of production attest to this new understanding of time. Moreover, the idea of society as an efficient machine would transform the way in which people approached modernization itself. Simmel's reading of society as an entity subjected to capitalism and massification is pertinent in this context. For him, as anthropologist Aram Yengoyan discusses, "What modernity creates are new forms of mediation which re-configure fundamental forms of society."[34] As a consequence, it has a particular impact on what Simmel calls "the mental life."[35] In the plane of industrial expansion there is a sensory engagement (sight, hearing, and smell) in all social interactions, as well as clear forms of cultural differentiation related to class, gender, location, and embodiment. Since sensorial perception is both objective and subjective, notions of time, distance, and space were constantly renegotiated in the interaction between the physical aspects of modernization and their mental representation.

The "assault on the senses" can also be understood within the parameters of that which archaeologist Nicole Boivin denominates "sensuality of the object": "The material world . . . evoke[s] experiences that lie beyond the verbal, beyond the conceptual, and beyond even the conscious . . . [objects] do not necessarily symbolise anything else: their very power may lie in the fact that they are part of the realm of the sensual, of experience, and of emotion, rather than a world of concepts, codes, and meaning."[36] Once again, the railroad provides a good example of one industrial space where this kind of experience became palpable. In his enlightening analysis of mass-culture individualization, Michel de Certeau briefly examines some of the effects that the railroad had on the way society was imagined. De Certeau sees the train not just as a revolutionary development in transportation but also as a disciplining space.[37] A train car can be understood as a classic Benthamian panopticon, for it limits the passengers' freedom by confining them to small spaces with no room for privacy.[38] In this "imprisonment," only sleeping or looking out the window could provide escape from other passengers' surveillance. The nature of this incarceration, to be sure, highlights the importance of the gaze and, in general, the sensory impact caused by the presence of the *other*.

This reorganization of the senses had a particular impact on the configuration of social foundries, especially the ones included in train travel accounts. The perception of confinement in the passenger car produced an effect of discontinuity and isolation that contributed to relativizing the narrative notions of time and space. For the early train travelers, one of the trip's novelties consisted precisely in adjusting the senses to deal with unusual perceptions of reality: "In apprehension of acceleration, the passenger's body came to signify the collision of reality and illusion, placement and displacement, transport and stasis, as a new disembodied vision complicated and transformed the sense of sight."[39] Perhaps the best way to see how this sensorial dislocation operated at the allegorical level is to analyze accounts of extraordinary events. In one of his travel chronicles, novelist Pedro Antonio de Alarcón describes his experience in a train accident. Testimony of one of the first significant collisions and derailments that occurred in Spain, Alarcón's account acquires epic tones by incorporating elements such as simultaneity or alternation in an attempt to capture the magnitude of the disaster:

> Del ténder y de la locomotora, que iban delante de mí llenos de gente, no se veía ya nada, sino humo, polvo, fuego; agua que corría de la caldera; las ruedas vueltas hacia arriba; las peñas saltando al empuje de la máquina, que aún quería andar después de haber encallado en ellas; algún hombre que se levantaba ensangrentado de debajo de aquellas destrozadas moles, dando alaridos; y nuestro vagón, al cual le tocaba volcar en seguida, y al que le faltaba poco para acabar de dar la vuelta ó para saltar en astillas.[40]

> (From the tender and locomotive, which were in front of me filled with people, there was nothing to see but smoke, dust, fire and water running out of the boiler; the wheels upturned; rocks jumping at the impact of the machine, which still wanted to move after crashing against the mountains; a man getting up under those bulks covered in blood and yelling; and our car, which was next to overturn, and that was about to complete the turn or jump in splinters.)

Alarcón's narrative markers underscore the inability of language to re-create the scene he was witnessing. He adds: "Nuestro dolor al ver muerto al eminente ingeniero Alfredo Lee, y en tan grave situación a su hermano; nuestro asombro al encontrarnos vivos; nuestro reconocimiento a Dios que nos había librado; el terror del pueblo que nos cercaba; los penosos cinco cuartos de hora que se tardo en sacar a Morlando Lee de debajo de la máquina, son cosas que no acertaría a describir." (Our pain at seeing eminent engineer Alfredo Lee's death, and his brother in such a grave condition; our amazement of being alive; our acknowledgment to God for saving us; the terror of people around us; the awful five-quarters of an hour that were necessary to rescue Morlando Lee from under the machine, all are indescribable things).[41] The image of the accident is fragmented and presented through the use of nominal clauses to highlight the difficulty of a narration that includes a profusion of simultaneous events. By separating the actual timing of the events from what happens in it, Alarcón manages to capture the way in which the collision distorts time: "¡Oh! fueron cuatro segundos, . . . pero cuatro inmensidades de pensamientos, de recuerdos, de angustias" (Alas! it lasted four seconds, . . . but they were four long thoughts, memories, angsts).[42] The same disproportion materializes at the narration's formal structure, where the account of a four-second episode extends for pages. Through this strategy, Alarcón translates his own experience so that readers can have an accurate image of events which few people had lived before.

Discursive negotiations of technological change that show the "assault on the senses" of modernization not only stemmed from exposure to physical or mechanical conditions; they also arose from experiences of socialization and personal interaction. Benito Pérez Galdós's novella *La novela en el tranvía* (A tramway story) (1871) provides a good example of how the streetcar's confinement and the close contact with other passengers could dislocate the perception of reality. Exploring a sort of technological imagination, Galdós's text criticizes the popular genre of serialized novels, a minor form of literature that in his view was unable to properly reflect society. The main character in *La novela* is the reader of an implausible and melodramatic story, an incomplete episode of which he inadvertently finds in his ride. His exuberant imagination leads him to think that the story he is reading not only is real, but also that some of its characters happen to be traveling in the cable car with him. Galdós's description of the unusual, yet unwelcome opportunity of scrutinizing others suggests the emergence of a new and rich fictional landscape:

> Recorriendo con la vista el interior del coche, examiné uno por uno a mis compañeros de viaje. ¡Cuán distintas caras y cuán diversas expresiones! Unos parecen no inquietarse ni lo más mínimo de los que van a su lado; otros pasan revista al corrillo con impertinente curiosidad; unos están alegres, otros tristes, aquél bosteza, el de más allá ríe, y a pesar de la brevedad del trayecto, no hay uno que no desee terminarlo pronto. Pues entre los mil fastidios de la existencia, ninguno aventaja al que consiste en estar una docena de personas mirándose las caras sin decirse palabra, y contándose recíprocamente sus arrugas, sus lunares, y este o el otro accidente observado en el rostro o en la ropa.[43]

> (Going over the interior of the coach, I examined one by one all my companions. What a diversity of expressions and different faces! Some seem to be completely indifferent about the people beside them; others evaluate the crowd with impertinent curiosity; others are happy, others are sad, that one over there yawns, the one a little bit farther laughs, and in spite of the shortness of the trip there is no one who would not want it to conclude sooner. Because among the thousands of hassles of existence, no other is more annoying than spending time with a dozen people, looking at each other's faces without saying anything, and reciprocally taking account of each other's wrinkles, moles, and this or that mishap on the face or clothes.)

In the story, in fact, the space of social interaction turns into a new narrative territory wherein reality and fiction converge to highlight the fragility of perception and the cognitive impact of urban modernization.

Writing, however, was not the only space in which codes and signifiers of progress were articulated to produce novel and insightful assessments of reality. Other discursive practices and artistic expressions also became sites for the negotiation of new material, cognitive, or sensorial experiences. In painting, for example, as cultural historian Stephen Kern suggests in his analysis of early twentieth-century art, representations of the malleability of time and space were a common motif.[44] Music, photography, and, of course, the newly created cinematograph provided suggestive ways to show the impact of industry and technology on human experience.[45] In Spain, these artistic expressions also constituted reflections on the ideas of nation and national identity that evinced the peculiarities of the country's modernization. Cultural critic Stephanie Sieburth, for example, has noted how appropriations of massification and industry were contingent upon the ideological polarization that characterized the late nineteenth century. For her, the response of writers and intellectuals to the social transformations of industrialization was very conflictive and even contradictory, as it not only confronted political ideas regarding past and tradition but also notions of high and low culture associated with phenomena such as mass consumption, large-scale manufacturing, and technological change.[46] However, cultural production transcended and complicated this dichotomic perspective by incorporating material aspects that molded the way in which different social groups perceived and appropriated science and technology and ultimately articulated their social foundries.

The Physics of the Social Foundry

As an abstraction of the steel industry, the social foundry operates within the theoretical framework provided by two major changes in the conceptualization of physical phenomena—the theory of the *field* proposed by Michael Faraday and developments in the study of thermodynamics presented by James Joule. During the nineteenth century, these advances converged to produce new ways of thinking the social construct. In the first case, the recognition of the existence of forces that were invisible but nonetheless had a perceptible effect on the physical world (magnetism, electricity) revealed that reality consisted not only of discrete objects and their locations in space but also of underlying conditions whose interaction determined the very existence of those objects and the way in which they were distributed. This meant that there could be no objective position from which to observe reality since its true complexity escaped the senses.[47] In the second instance, the theorization of physical models for the dissipation of heat in a mechanical system suggested that absolute essences had ceased to define the natural world, which was now understood as a set of dynamic elements that manifested themselves as energy when they came in contact with their surroundings. This mutability led to the basic principle that there is a finite amount of energy in the universe that surfaces in multiple forms. The study of natural phenomena under these premises required the institution of universal premises. The first of these, or the first law of thermodynamics, establishes that energy can neither be created nor destroyed; the second proves that all processes of transformation give off energy in the form of heat. This continual outlay of energy and its consequences in the production of work is what is known as entropy, a term introduced by German physicist Rudolf Clausius in 1865.[48] On the basis of these principles—and given that the majority of technical and industrial advances from this period depended on the manipulation of heat—thermodynamics and its laws ended up permeating the collective imagination. Since many politicians, writers, and intellectuals of this period were also scientists and educators, their comprehension of these concepts allowed them to take a theory about physical phenomena and turn it into a tool for evaluating society.

As these recently discovered physical principles seemed to offer plausible explanations for natural and social phenomena that had been repeatedly justified on the basis of faith, their popularization in Spain intensified the antagonism between religion and science. As Pohl-Valero explains, interpretations of reality based on the laws of thermodynamics openly questioned the Catholic dogma and the Church's authority:

> Dentro de este proceso de reformas políticas y sociales, la conservación de la energía, junto con la teoría de la evolución, fue una teoría científica muy importante a la hora de ofrecer una visión del mundo que se regía exclusivamente por leyes naturales y donde la mano de Dios y lo sobrenatural no tenían cabida. En este sentido, la termodinámica articuló discursos para cuestionar el poder de la Iglesia y para proponer una nueva moral basada en la razón.[49]
>
> (Within this process of political and social reformation, the law of conservation of energy, along with the theory of evolution, was a fundamental scientific tool that offered a particular world vision wherein the natural laws regulated everything, and where God or the supernatural did not have a place. In that sense, thermodynamics allowed the articulation of ideas that questioned the Church's power, and proposed a new morality based on reason.)

Tying social control and state politics to an industrial conception of the country was a natural extension of this logic, in which science gave new meaning to the body–machine and society–industry analogies. These codifications of reality transcended political agendas and religious dogmatisms in such a way that not only liberal scientists and educators but also thinkers with conservative ideologies appropriated scientific and industrial imagery to articulate their proposals and advance their agenda for the country's modernization.

The discussion about the role of the working classes within these projections of society provides a good example of how different ideologies resorted to physics, and particularly to thermodynamics, to interpret or justify social change. In Spain, the need for a rapid expansion of the industrial base demanded the implementation of labor-intensive apparatuses of production and the strengthening of the hierarchical social stratification. It was evident, as Jo Labanyi aptly suggests, that in this particular environment labor and capital could not band together, even when their clash contradicted some of the premises on which liberalism was substantiated.[50] To validate this exploitative model, the worker needed to be understood as one more cog in the machinery of industry, an object subjected to the same laws of physics that governed technical progress. In fact, as Pohl-Valero highlights, concepts and terms associated with thermodynamics were used in explaining the transformation of work into capital.[51] Such was the case of notions like labor power (Arbeitskraft), a term coined by German physicist and physiologist Hermann von Helmholtz and later implemented in Karl Marx's theorization of capital. According to Helmholtz, the body was the site where all the energy that sustained labor was converted, through processes that followed physical laws, into the necessary dynamism for the operation of factories and other industrial centers.[52]

The two principles governing this transformation—the conservation of energy and the dissipation of heat—therefore provided grounds for the use, and in many cases the abuse, of workers. To avoid the loss of dynamism due to the dissipation of energy, as predicted by the second law of thermodynamics, for example, workers were required to unceasingly inject their labor force into the system. On the other hand, the conversion of this investment into capital was only possible through an unequal distribution of resources that, as anticipated by the first law, guaranteed the permanent circulation of energy. The ways in which effort and capital were allocated within the industrial economy following these principles thus became relevant for explaining factors such as the division of labor and the concentration of wealth. The utilitarian and materialist nature of this rhetoric, however, soon came into conflict with the moral and religious principles of many politicians and entrepreneurs—an issue that became even more notorious after the popularization of the work *Kraft und Stoff* (1855), written by German physician Ludwig Büchner and translated into Spanish in 1868 under the title *Fuerza y materia: Estudios populares de historia y filosofía* (*Force and Matter: Empirico-Philosophical Studies, Intelligibly Rendered*). The text's materialist interpretation of the natural world produced a heated debate among Spanish intellectuals. According to Büchner, the first law of thermodynamics was evidence of a cyclical and eternal universe that did not rely on divine intervention to function: "La tierra y el universo son eternos, porque la eternidad es una cualidad *inherente á la materia*. Pero el mundo está sometido á modificaciones, y por eso el hombre cuyo espíritu no ha iluminado aún la ciencia cree que ese mundo es también limitado y pasajero" (The Earth and Universe are eternal, because eternity is an *inherent quality of matter*. But the world is susceptible to change and that is why men, whose spirit has not yet been illuminated by science, believe that such a world is also limited and temporary).[53] Notwithstanding its implications, in the striving for modernization Catholic thinkers received Büchner's ideas with certain enthusiasm, though they were always careful of adapting their scientific foundations to suit less materialistic ideologies.

As Pohl-Valero indicates in his studies on the political uses of physics in nineteenth-century Spain, the most prominent public figures in the debate elicited by *Fuerza y materia* were educators and scientists Enrique Serrano, José Rodríguez Mourelo, and Laureano Calderón.[54] Their particular use of thermodynamics in the articulation of educational and social plans is worth noting in this context. Serrano, for example, developed an educational program based on the importance of physics, particularly on the study of energy. His attempt to bring together different scientific fields through the examination of thermodynamics, as stated in his 1874 article "La evolución en la naturaleza" (Evolution as part of nature), also touched upon the country's need for social reformation, establishing a connection between institutional progress and a mechanistic view of society. In Serrano's view, dynamism could be understood to be the result of the constant transformation of energy.[55] His notion of a universe in constant evolution indicates a teleology grounded in the first and second laws of thermodynamics, according to which all matter (organic and inorganic) was animated by an internal generative force that could be defined in metaphysical terms as God.[56] Without forgetting human intervention, Serrano thus assigned an ontological, rather than a natural, character to this transformation. Similarly, for Rodríguez Mourelo nature was a living organism whose progress depended on the continuous transformation of energy initiated and sustained by the power of God, but actually brought about by biological entities.[57] By understanding science as a combination of material and spiritual elements, both scientists proposed transformation as a structural foundation of reality in terms similar to those found in Büchner's text. Rodríguez Mourelo's 1882 article, "Concepto de la energía" (The concept of energy), for example, suggests that considering progress as a natural and divine manifestation required all the same a scientific conception of the universe:

> Suponed conocida la masa del sol, y determinada la velocidad de sus acciones, aplicando la fórmula de la fuerza viva, pronto diríais cuánto había de durar ese sol y qué masa y qué acciones se necesitarían para formar otro que diese tanto calor, tanta luz y tanta vida.
>
> Aún hay más: suponed conocido esto mismo en una sociedad; ¿no podréis juzgar al momento de su estado de cultura, civilización y progreso? ¿no podréis apreciar sus males y sus necesidades para acudir á ellos y remediarlos, y en cuanto fuere posible, regenerar aquella sociedad pór virtud de sus mismas fuerzas vivas convenientemente utilizadas y repuestas?[58]

(Suppose that one knows the sun's mass, and that the speed of its movements has been determined; by applying the formula of the living force, then you could tell how long the sun would last and what mass and which actions would be needed to form another one that will be able to produce as much heat, light, and life.

And there is more: suppose that one knows all these variables in a particular society; would you not be able to instantly determine the state of its culture, civilization, and progress? Would you not be able to assess its problems and needs with the purpose of going over them and solving them, and, if possible, regenerating that society by conveniently restoring and using its own strengths?)

By connecting physics and social diagnosis, this passage successfully combines scientific materialism and religious spirituality. Such a search for ideological balance directly relates to Krausism, a doctrine to which Serrano and Rodríguez Mourelo subscribed, and of which authors such as Laureano Calderón and José Echegaray were influential adherents.

In fact, Echegaray's approach to the assessment of labor power provides another excellent example of this interest in balancing materialism and spirituality. In his talk "Aplicación de las fuerzas naturales a la industria y al comercio" (Application of natural forces to industry and commerce), part of a series of readings given at the Círculo de la Unión Mercantil de Madrid (Circle of Commercial Alliances of Madrid) between 1879 and 1880, he explains,

> La economía política distingue dos clases de industrias y divide el trabajo humano en dos grandes grupos: industrias, á que dá el nombre de materiales, y cuyos productos son materiales también, é industrias que puedo designar con la denominación de inmateriales, porque sus productos, sus creaciones, son, digámoslo así, del órden espiritual. Unas y otras, dentro de la economía política, están sometidas á las mismas leyes.[59]
>
> (Political economy discerns two kinds of industry and divides human labor into two big groups: industries that can be qualified as material since their products and creations are indeed material, and industries that I will designate as immaterial because their creations are, to be sure, of the spiritual order. Within the realm of political economy, both obey the same laws.)

At the convergence of the scientific world and the "spiritual order," human factors and social components played a decisive role. As he points out later in the same speech, the link between material and immaterial industries was indeed human labor:

> Si yo os pregunto: "¿qué analogía hay entre todas estas industrias, qué hay de común entre todas ellas, qué es lo que en todas ellas se repite de la misma

> manera, y cuál es la semejanza que puede existir entre esta multiplicidad de trabajos humanos?" solo al hacer esta pregunta, doy ya por resuelto el problema. Porque al decir: "¿qué factor común hay entre los diversos trabajos humanos?" ya lo he dicho: unos y otros, y todos ellos, tienen de común una cosa; *el ser trabajo humano*, y esta palabra *trabajo* y la idea que representa se repiten constantemente para todos ellos.[60]
>
> (If I ask you: "What analogy is there among all these industries? What is common to all of them? What element is repeated in all of them in the same way? And what are the possible similarities within this multiplicity of human activities? Just by posing this question, I am giving away the answer. Because by asking "what is the common factor among the diverse human activities?" I have already said it: one or the other, and all of them, have one thing in common; they are *human labor*, and the word *labor* here, and the idea it represents, is a constant shared by all of them.)

In this image, human labor is mediated by productivity, that is, by the constant need to transform the investment of energy in palpable results. Yet yielding this outcome was not possible without an adequate allocation of labor forces and a precise organization of society.

Political, economic, and religious constraints clearly delineated the interaction between industrial modernization (in both its material and intangible aspects) and the way in which authors understood society. Within the material field of industrialization, the assimilation of scientific and technological ideas was intrinsically connected to ideological agendas. A review of the historical conditions that defined these perspectives is therefore crucial to understanding how the particular codifications and interpretations of reality that I have termed social foundries materialized.

Spanish Modernization and Its Discontents

Even when science, technology, and culture can be seen as self-contained, independent fields, they all are affected by sociohistorical conditions. As a matter of fact, cultural production and scientific discourses are, in their own way, representations of reality that respond to historical specificity. Take for example what a trivial event—the installation of the first Bessemer converter in Bilbao in 1862—says about fin-de-siglo Spain. The arrival of this technology to the country constituted a landmark in the efforts to modernize one of the most vigorous sectors of the economy, thus confirming the Basque region as one of the world leaders in steel production.[61] The innovative system, invented a few years earlier by Henry Bessemer, had drastically changed the business by reducing the elevated costs of manufacture. Normally, steel is made by combining specific proportions of iron and carbon; the main challenge in this process is obtaining the elevated temperatures needed to melt both elements. Additionally, neither iron nor carbon occur in pure form in nature, which causes residual contamination in the process of mixing them together. Steel's final quality depends precisely on the reduction of undesirable subproducts. Bessemer's great innovation consisted in using oxygen as part of the smelting process; the resulting reaction not only increased the mix's temperature, saving costs, but also eliminated residues. In order to properly work, however, the system required the use of a specific kind of iron.[62] Coincidentally, most of Spain's exploitable reserves of the metal had the chemical properties demanded by the Bessemer process; yet, and notwithstanding this unparalleled natural advantages, by the end of the nineteenth century the consolidation of the industrial economy continued being problematic. Many historical factors contributed to this particular conundrum.

Spain started off the nineteenth century on the wrong foot. Despite the moderate nature of Carlos IV's administration (1788–1808), the period was especially turbulent: the conflictive relations between the king's favorite, Manuel Godoy, and influential sectors of society had led the country into a moral crisis. While thinkers in the opposition were imprisoned or exiled, the Enlightenment's reforming principles and ideas got dismantled; consequently, a sense of disillusionment prevailed among intellectuals and artists. Particularly telling in this regard was the banishment and detention of Gaspar Melchor de Jovellanos in 1801, after unfruitfully fighting the government's rampant immorality from his official position as minister of religion and justice. Additionally, in 1805 Spain suffered a humiliating defeat in Trafalgar that served as a prelude to Napoleon's invasion in 1808. As mentioned earlier, it would be amid the struggles to expel the French invader that questions of identity and patriotism gained an unforeseen momentum; notions that were further complicated by the various campaigns for independence in many of Spain's colonies in America. In social and economic terms, during the first decades of the century the national debate focused either on defending the structures of the Old Regime or on promoting a progressive form of government based on people's sovereignty and the separation of Church and state. Both the attempt of Carlos's successor, Fernando VII, to establish an absolutist Catholic monarchy and the progressivists' sustained efforts to consolidate a liberal society, however, would partially fail in giving the country a new direction at this critical moment. As a matter of fact, the 1812 constitution had to be negotiated on the basis of these two perspectives' irreconcilable character.[63] The long reign of Fernando VII after his restoration to the throne in 1813 would confirm the historical weight of this confrontation by pushing a conservative agenda that strongly antagonized any notion of progress.

Fernando VII's reactionary rule gave extended privileges to the Church and the aristocracy, prerogatives that had severe repercussions on education, the promotion of science, and the implementation of technology. Thus, for instance, while by 1765 there were numerous organizations and institutions dedicated to the study of nature in Spain—the Botanical Garden, the Natural History Cabinet, and the Academy of Medicine, among others—by 1833 institutional efforts to advance science and technology had practically disappeared.[64] Instead of opening the country to the possibilities of modernization, the king's passing that year exacerbated the social tensions and ideological conflicts that had characterized his reign. One of the main sources of conflict were the changes in the laws of succession Fernando introduced so that his daughter, Isabel, could rule in place of the legitimate heir to the throne, the king's brother, Carlos. This dispute further divided the country, as Carlos's reclamations had the support of wide traditional sectors of society, and, ideologically, offered a continuation of the conservative principles that had characterized his brother's administration.[65] The reallocation of resources to fight the armed uprising of loyalists to Carlos would contribute even more to the retardation of industrial implementation.

In general terms, nonetheless, the monarchical succession had positive effects on the country's modernization. With a weakened apparatus of ideological regulation, previously exiled liberal thinkers and dissidents returned to the country. Many of these expatriates served as vehicles for new ideas, knowledge, and publications whose divulgation had been restricted on the peninsula. During the second half of the nineteenth century, the educational system also experienced several reforms that focused on securing curricular independence and promoting scientific and technological innovation.[66] Despite isolated efforts to modernize the state administration, however, it would not be until 1868, after the liberal revolution, that many of Fernando VII's problematic legacies were finally overcome. With the restoration of the monarchy in 1875—a setback for the democratic and progressive policies implemented in the short six-year revolutionary period—this momentum of liberal reformation came to an end, worsening the profound division between promoters of progress and defenders of tradition. At the same time, phenomena related to industrial modernization, such as the working-class unrest, the degradation of the environment, or the rapid urban expansion, complicated the negotiation in the cultural realm of the multiple changes of society.

In the political, literary, and scientific arenas, textual production evinced the convoluted "web of affinities"[67] that entangled science, technology, and culture at the end of the nineteenth century. As the level of industrialization in Spain came close to that of other European countries, the disparity between the practical and the theoretical applications of science became more evident. Even though sectors such as mining, steel, and textile production had consolidated, research in fields like physics, chemistry, or mathematics remained scant—thus, despite the wide implementation of scientific ideas, a national scientific tradition was not clearly defined.[68] Furthermore, a marked emphasis on economic profit over community engagement prevented the country from successfully linking research, technological development, and social improvement. The lack of effective reforms to balance science education in order to serve both practical and theoretical ends was problematic as well. From the early stages of modernization, educational systems in Europe were structured to serve the needs of industry while producing concrete scientific and technological advancements. Under that model, an important portion of industries' economic benefits were reinvested in educational institutions that worked on creating new methods to improve productivity.[69] In Spain, however, this cycle was not implemented until later in the twentieth century. As a matter of fact, during the second part of the nineteenth century the opposite was closer to reality: an insufficient investment in research and development made the creation of institutes of higher education unnecessary, and the absence of those, in turn, impeded a faster and more productive industrial development.[70]

Another factor that complicated the consolidation of a national scientific tradition was religion. Catholicism became a conflictive ideological element in the cultural negotiation of modernization. While ecclesiastic authorities radically condemned scientific theories and discoveries considered dangerous to Catholic dogma, promoters of progress adamantly attempted the refashioning of science so that it could be an instrument to celebrate the creations of God. Take, for example, the explanation of the purpose of science that José Echegaray offers in his inauguration speech at the Royal Academy of Sciences: "Que al fin es la ciencia, por más abstracta que en sus concepciones a primera vista parezca, germen fecundo de progreso para pueblos, enérgico purificador del alma, luz que alumbra a la humana inteligencia con divinos resplandores" (In the end science, no matter how abstract its conceptualization seems to be at first, is a fertile germ for the progress of peoples, a powerful soul purifier, the light that illuminates human intelligence with its divine radiance).[71] Echegaray uses the idea of God as the source of scientific development, turning adjectives that usually refer to the divine power—"fertile germ," "soul purifier," "light that illuminates"—into qualifiers of science. His symbolization is therefore informed by religious ideas: allusions to the soul and the light can indeed be associated in this context with some of the central principles of Catholicism. At the same time, he sees intelligence as a "divine radiance," positioning reasoning and scientific knowledge at the center of human progress.

Yet Echegaray's attempt to harmonize science and religion did not prevent him from criticizing the Church's negative role in the development of a national scientific tradition. Referring to the state of mathematics, in the same speech he adds, "Angustiosas reflexiones se agolpan a mi mente al recordar este nuestro lastimoso atraso, y atraso crónico, en uno de los ramos del saber que más glorias han dado a la época moderna, y que tanto contribuye a vigorizar las más nobles facultades del alma" (When I think of our sad, chronic backwardness in one of the branches of knowledge that has contributed the most to strengthen the soul's most noble capacities, distressing reflections pile up in my mind).[72] Scientific knowledge and religion are comparable in that they "strengthen the capacities of the soul." Contrary to what conservative advocates suggested, industrial modernization, technological development, and scientific research were not going to push the country away from tradition. Instead, the collectivizing rhetoric in Echegaray's speech confirmed the nationalistic dimension of the "past glories," a legacy of military and political success that had been guided by religion. However, since the colonial expansion in America, Spain had been incapable of producing any "discovery" or "conquest," a situation that only an adequate promotion of science could change:

> No puede, en verdad, gloriarse nuestro país de ningún importante descubrimiento, porque cuando tan rezagada queda una nación, harto hace con alcanzar a las que en tres siglos la aventajan; pero el porvenir es suyo, su voluntad será enérgica, el campo del saber es infinito, y genios tendrá cuando libre de fatales trabas, y conquistada la libertad filosófica, que es la libertad del pensamiento, se lance de lleno al estudio de esta gran ciencia que dió a Descartes, a Newton y a Leibniz nombre inmortal.[73]

> (Our country certainly cannot brag of any important discovery, because when a nation is so left behind, it does not achieve much in catching up with the ones already ahead by three centuries; but the future is hers, her determination will be strong, the field of knowledge is infinite, and Spain will have geniuses when, released from prejudice and able to conquer philosophical freedom, it leaps into the study of the science that gives Descartes, Newton and Leibniz their immortality.)

Echegaray's denunciation of religion as an evident obstacle to scientific progress in Spain generated a heated debate about the historical role that Catholicism had played in the country's modernization. The discussion, part of what is known as *Polémica de la ciencia española* (polemic over Spanish science), would expose once again the ideological conflicts that marked the relationship between science and society in fin-de-siglo Spain.[74]

The problematic educational reforms enacted during the last portion of the century also reveal the conflictive nature of the different political approaches to scientific modernization. In 1875, for example, in a clear gesture of support of government's conservative values, and aligned with a political ideology known as *integrismo*,[75] secretary of education Manuel Orovio withdrew the *libertad de cátedra*—a mechanism designed to prevent the government's interference with the contents, teaching methodologies, and organization of higher education. The enforcement of this decree compelled many intellectuals to abandon their university posts and elicited, as pointed out earlier, the creation of the Institución Libre de Enseñanza. Under the Church's strict supervision, scientific instruction at the university level limited its scope to serve the country's immediate needs.

As a consequence, and in order to continue consolidating the industrial apparatus, it was inevitable to import knowledge and technology from other countries, a process that not only contributed to the delay in the creation of a national scientific tradition but also justified the need for foreign investments in the local economy.[76]

The expansion of industry in specific areas of the country, moreover, provoked the relocation of large portions of the population, dramatically changing the national demographic distribution and challenging the pretended homogeneity of national identity. In fact, one of the most important consequences of these intangible (technology, ideas, knowledge) and physical displacements (population) was the strengthening of nationalistic sentiments, a phenomenon that may be explained as a reaction against the circulation of foreign paradigms and the disruption of traditional social orders. The case of the Basque Country was emblematic in this regard. There, as political scientist M. K. Flynn explains, industrial modernization polarized society into two distinctive groups—the working class and the industrial oligarchy—whose particular interests were at odds with the local culture: "In the case of Basque nationalism, popular support developed in relation predominantly to neither a hierarchical cultural division of labor . . . nor a segmental cultural division. . . . Rather it developed from a partial realization of both in which nationalists neither identified themselves with the top of the socio-economic system dominated by hispanicized Basque oligarchs nor, while tending to come from sectors important in the overall constitution of regional society, occupied high-status positions."[77] Due to the industrial boom, large groups of workers from different provinces relocated in the region. In most cases these individuals had no interest in assimilating to the local cultures. Basque industrialists, for their part, pledged their allegiance to the monarchy and the central government and were not keen on embracing the regional identity. Values directly associated with that which was considered the Basque character were thus socially marginalized, pushing the urban middle classes into nationalism as the only way to defend their own status. Similar processes took place in other industrial hubs, where not only internal migration but also international interference had an effect on the consolidation of nationalistic views.

Subordination to the control of other European nations over the local economies stressed the evident inconsistencies between the country's self-image of modernity and its reality. In that context, many scientists, intellectuals, and politicians challenged Spain's alleged exceptional geographical, ethnic, linguistic, and religious condition. The military defeat in the Spanish-American War of 1898 confirmed this criticism and created new opportunities to question the status of national modernization. The *disaster*, as this defeat would later be known, complicated the tensions between liberals and conservatives. Blaming the country's decline on excessive materialism, both sectors considered that Spain could be rescued from its social, economic, and moral afflictions by strengthening its traditional values while embracing modernity's capacity for social transformation.[78] Even though authors interpreted the country's decline in subtly different ways, in general all concurred in the need for reallocating public resources to improve higher education and in the urgency of consolidating a productive industrial society to be on a par with other European nations.[79] As a consequence, they repudiated the institutional structures of the Restoration and rejected the traditional political class, whom they blamed for the country's state of backwardness and decay.

Industrialization became a powerful source of inspiration—authors used industrial referents to reconsider the country's problems, the strategies to address them, and the possible scenarios for the future. By confronting different ideologies, scientific and technological change configured a material field wherein new modes of perception, conceptual apparatuses, and allegoric and metaphorical interpretations helped articulate and carry out the projects of national modernization. The evaluations derived from these processes of cognitive appropriation deal with clearly recognizable issues: social-class mobility, science education, tensions between science and religion, and environmental cost are some of the most important ones. At the discursive and textual levels, these interpretations of national reality were expressed following well-established genres and rhetorical structures, including fictional accounts, scientific treatises, travel chronicles, popularization articles, and academic lectures. The analyses I undertake in the following pages center first on identifying the novel mechanisms of representation that authors used in their assessments of the country, and then on exploring their suggestions and proposals. In other words, I study how their social foundries problematized ideas of progress, national modernization, and identity.

CHAPTER 2

Economy and Other Matters of State

By the end of the nineteenth century, the industrial exploitation of natural resources had completely transformed the country's economy. In that context, knowing and understanding the peninsula's physical dimensions and wealth were important not only for envisioning economic strategies that guaranteed prosperity but also for symbolically reaffirming notions of superiority and uniqueness that linked national identity to the particularities of its geological and geographical situation. The idea that Spain's meridional location gave it an advantage over its European counterparts, securing, for example, an abundance of resources, had to some extent discouraged a conscientious study of the land. It comes as no surprise, then, that the industrial boom and its demand for minerals had finally pushed the government to generate an accurate portrait of the country's geological wealth. Under the title of *Mapa geológico de España* (Geological map of Spain), this project gathered the most prominent geographers and geologists of the period, establishing a dialogue between science, industrial development, and society that would shape the ways in which national modernization was projected.[1] In fact, the data compiled in the study were later used in economic, social, and political analyses addressed to specialized and not-specialized audiences. Some authors, for example, incorporated this information into their criticism of the state's incompetence in developing an independent, self-sufficient industrial operation, proposing that the country's natural wealth was of no use without adequate means of scientific production and technological implementation.

This new understanding of the peninsula's natural resources, and of their actual potential to satisfy local and international demands, also complicated the national debate on economic protectionism. Within this context, concepts such as conservation, dissipation, and exhaustion became common referents in the analysis of the forces that regulated the economy. This was not a coincidence, considering that the physics of energy, particularly the principles that govern heat and its transformation into work and movement, were being widely used to explain social phenomena. This view of the economy, in which physical principles informed the interpretation of reality, moreover, seemed to urge the enforcement of conservative measures. As Bruce Clarke notes, "Nature's administration disallows deficit spending," and science proved the "fiscal constraints" of social and economic forces.[2] In the same way that energy or power needed to be budgeted to guarantee the expected outcomes in industrial production, social variables had to be scientifically calculated to measure, understand, and accurately predict society's development.

This chapter discusses mainly the ideas of three authors: scientist and politician José Echegaray, geologist Lucas Mallada, and economist and legal philosopher Pere Estasen i Cortada. Through the use of physical, geographical, and geological referents linked to industrial and economic modernization, in their diagnoses of what they saw as an ill national body, these social scientists explore many of the interpretative possibilities offered by the cognitive and material transformations of progress. By studying the unchangeable conditions of the land and the uneven forces of the economy, they question the geographical determinism and national exceptionalism that justified ideas such as the immutability of the Spanish character or the country's supposed predestination to fail. The works analyzed here advocate for the country to reformulate the past in order to confront the social challenges of the present. To achieve this goal, a new understanding of the national potential was necessary before any attempt could be made to enact a plan of economic reform that balanced the domestic and international exchange of goods, human capital, and knowledge. Such is the essence of the *social foundries* advanced by Echegaray, Mallada, and Estasen, interpretations of society that emphasize the agency of peasants and workers, and most importantly of scientists and educators in transforming the country's symbolic wealth into the natural and material resources required for modernization.

José Echegaray's Identity Collectivism

Although José Echegaray's monumental oeuvre covers a myriad of fields, including physics, mathematics, economics, and engineering, his work can be grouped in three big categories: education, politics, and literature. After graduating in 1854 from the Escuela de Ingenieros de Caminos in Madrid (School of Roads Engineering), Echegaray joined the faculty as instructor of mathematics and physics. In the mid-nineteenth century, the school was going through a period of renovation, pushed by the liberal ideologies that had materialized after Fernando VII's demise in 1833. Consistent with its central mission of serving as a pillar of the country's industrial modernization, the school's curricula mainly focused on teaching practical aspects of science. Echegaray, nonetheless, like many other scientists and educators of the time, considered theoretical and practical knowledge equally essential for the development of any modern nation, as he would confirm years later in his celebrated inaugural speech at the Royal Academy of Sciences in 1866:

> Así, señores, es la ciencia eminentemente útil, no de una manera indirecta por sus aplicaciones, sino directa e inmediata, porque directa e inmediatamente y por su propia virtud, satisface altísimas necesidades humanas, y del mismo modo que el cuerpo busca el pan de cada día, busca el alma, hambrienta de belleza y de verdad, algo que satisfaga las aspiraciones a lo infinito de su inmortal esencia; busca la verdad, repito, por esa misteriosa atracción que entre la verdad y el pensamiento existe, y que hace que la razón vaya tras ella anhelante, y sin ella muera, y con ella viva, y que al hallarla en su esencia divina se sumerja y se bañe gozosa como en océano de luz.[3]
>
> (It is then a fact, gentlemen, that science's usefulness is not indirectly the result of its applications, but directly and immediately the consequence of its own virtue, because it directly and immediately satisfies complex and demanding human needs. And in the same way that the body needs bread every day, the soul, hungry for beauty and truth, needs something that satisfies its aspirations of transcendental immortality; it needs to reach the truth, I repeat, because of the mysterious attraction that exists between truth and reasoning, the same force that justifies the eager pursuit of truth, without which reason dies but with which its divine essence lives immersed in an ocean of light.)

For Echegaray "virtue" and "truth" were essential features of scientific activity. The humanist tone of the passage thus highlights the place that science and reasoning had in a new spirituality, one separated from the Church, but at the same time universal and individual. Moreover, both pure and applied sciences, as mirrors of truth and reasoning, balanced the material and spiritual aspects of a modernizing project that needed to resonate with the peculiarities of national identity. During his time as a scholar, Echegaray went back to this idea time and again by publishing numerous studies, academic articles, and texts of scientific dissemination in which he not only presents and discusses the theoretical principles of mathematical or physical ideas but also provides ample examples of their concrete application to solving practical problems.[4]

After a short period dedicated to teaching, Echegaray moved to enlarge his circle of influence by entering the political arena. Serving in some of the most important positions in the country's administration, he had the opportunity to offer elaborated and extensive diagnoses of the national situation. Echegaray's political agenda focused on points such as freedom of speech, economic liberalization, and higher education. For him, the state should not interfere with the economy or impose ideological restrictions on education. These political convictions explain his active participation in organizations such as the Sociedad Libre de Economía Política (Free Society of Political Economics), founded by economist Laureano Figuerola i Ballester with the main goal of defending and promoting ideas and doctrines related to economic liberalism. Additionally, during the six years of revolutionary governments and the outset of the Bourbon restoration, Echegaray served as general director of public infrastructure (1868–69), minister of development (1869–70 and 1872), and secretary of finances (1872–73), and formed part of the Congress of Deputies (1876).

Echegaray's last area of influence was literature. After being a very active politician, once the administrative system put in place by the architects of the Restoration confirmed the failure of the liberal revolution, he abandoned the public arena. Echegaray then devoted his life to writing, focusing his artistic endeavors on renovating the theater, a genre that had lost its vitality throughout the nineteenth century in spite of its celebrated history in the country. Echegaray published uninterruptedly from 1874 until his death in 1916, leaving a prolific legacy of literary creativity that served as a referent in the consolidation of the genre during the twentieth century.[5] Because of the relevance of this work, and considering his essential role in the academic and cultural spheres, in 1904 Echegaray was awarded the Nobel Prize in Literature, a contentious honor that in the view of his critics rewarded more his trajectory as one of the main protagonists in the history of Spain than the artistic quality of his work.[6]

In order to better understand Echegaray's privileged position within the material field of industrialization and his role as a promoter of economic liberalism, in what follows I analyze in detail his lecture "Aplicación de las fuerzas naturales á la industria y al comercio." In this talk, part of a cycle of conferences offered at the Círculo de la Unión Mercantil between 1879 and 1880, Echegaray combines scientific, technological, and industrial imagery to accurately evaluate social and economic models that could facilitate modernization without disfiguring national identity. To support my analysis, I also turn to other speeches and public interventions, such as the already alluded speech at the Royal Academy of Sciences or the talk he gave at his inauguration as a member of the Real Academia Española in 1894. Not only does the production of these texts coincide with the three fronts—educational, political, literary—on which Echegaray stood out during his life as a public figure, but their content exposes the most important aspects of his social foundry, which includes a progressivist program of economic modernization.

In the previous chapter, I provided some historical clues to contextualize the *Polémica de la ciencia española*, a public debate that intensified during the last part of the nineteenth century around the perceived critical condition of science, scientific research, and technological development in the country. Confronting different perspectives on the possible causes for this situation, this discussion polarized the country around the question of modernization. It was precisely

Echegaray's speech in his inauguration as member of the Royal Academy of Sciences in 1866 that fueled this dispute. Bringing this speech back into my analysis is also useful to underscore some of the central elements in his particular appropriation of scientific and industrial referents to evaluate the country's problems. From its title, "Historia de las matemáticas puras en nuestra España" (History of pure mathematics in our Spain), the talk sets out a series of contrasts underpinning the system of counterweights on which the country's national ethos had been articulated. By pointing to the inward-looking notion of Spain's exceptionalism, the possessive adjective—"our"—also delineates scientific labor as a collective endeavor. Embodied by mathematics, this space of intellectual reflection, ingenuity, and creativity fulfilled a primordial role in the process of consolidating identity. For science actively intervened in the most essential and, at the same time universal, aspects of human nature, the nation could revitalize its essence by encouraging its practice. Echegaray reframes the incompatibility between tradition and progress by contrasting the history of mathematics in Spain to that of other European countries, and concludes that the local backwardness originated during the sixteenth century, when the religious tensions between Catholicism and Protestantism deviated the course of scientific and technological development.

For Echegaray it was precisely in the context of the sixteenth- and seventeenth-century religious dogmatism that the progress of mathematics in Spain stagnated, becoming in some way the negative reflection of the advances made in other places with less severe regimes of ideological restriction. As evidence for his argument, Echegaray pointed to Nicolas Antonio's bibliographical catalog *Bibliotheca Hispana*, in which the celebrated bibliographer (1617–1684) had compiled information of every work published to date on the peninsula:

> Abro la Biblioteca hispana, de don Nicolás Antonio, y en el índice de los dos últimos tomos, que comprenden del año 1500 al 1700 próximamente, tras muchas hojas llenas de títulos de libros teológicos y de místicas disertaciones sobre casos de conciencia, hallo al fin una página, una sólo, y página menguada, que a tener vida, de vergüenza se enrojecería, como de vergüenza y de despecho se enrojece la frente del que, murmurando todavía los nombres de Fermat, de Descartes, de Newton, de Leibniz, busca allí algo grande que admirar, y sólo halla libros de cuentas y geometrías de sastres.[7]

> (I opened *Bibliotheca Hispana* by Don Nicolás Antonio, and in the Table of Contents of the last two volumes, the ones comprising the years from 1500 to 1700 approximately, after many pages filled with theological titles and mystical dissertations on cases of contrition, at last I found one page, and only one, a diminished page that if alive would have blushed, as the reader would have done if, with shame and spite, and still murmuring the names of Fermat, Descartes, Newton, Leibniz, was hoping to find something to

admire there, but instead only finds titles of accounting books and tailor's geometries.)

Careful not to cross the limits of heterodoxy or materialism,[8] in his assessment Echegaray makes religiosity partially responsible for the deplorable situation of mathematics and, by extension, for the country's scientific, technological, and industrial backwardness. Referencing the inquisition, Echegaray highlights the repressive role played by the Church in the past and hopes for a future in which religion is no longer an obstacle to modernization: "Contra algún obstáculo se habrá estrellado sin duda el genio nacional, e importa mucho conocerlo, repito, para evitarlo en lo futuro, si ya desapareció, para acabar de destruirlo, si aún quedan restos; para que llegue al fin el día en que se borre la mancha que en el siglo XVII, siglo del despotismo y de la intolerancia, cayó sobre nuestra historia" (National genius had to have encountered a serious obstacle, and, I repeat, it is important that we understand it so we can avoid it in the future, if it is already gone, or complete its destruction, if some of its remains prevail; so the day arrives in which the stain of despotism and intolerance that the seventeenth century produced in our history disappears).[9] From this perspective, the hurdles of national modernization were the direct result of historical processes to which national identity was inextricably linked.[10]

By presenting the history of mathematics as an economy of scarcity, this metaphoric relocation of identity highlights the intimate causes of Spanish underdevelopment: "Amarga, tristísima verdad, bien lo conozco y lo siento, pero gran verdad también, y fuerza es repetirla para que perdamos ilusiones halagüeñas, que sólo pueden servir para hacer mayor el daño" (A bitter, tragic reality, I admit it and I am sorry, but an important reality no matter what, and it is essential to say it many times so we reject flattering delusions, misconceptions that are only good to worsen the damage).[11] Framed within a sort of medical language, Echegaray diagnosis emphasized the need to dimension the proportion of the country's malady; a task that required a critical evaluation of the national character to reject any "misconception" of greatness and exceptionality.

While the address at the Royal Academy of Sciences established the basis from which to contrast the errors of the past and the opportunities of the present, offering clues about the nature of the national afflictions, its scope prevented it from being a complete analysis of other mechanisms and forces that operated in society. A more complex reading of the social machinery using the lenses of physical laws, industrial imagery, and technical concepts is undertaken in other interventions that look at the importance of work and social cooperation in the regeneration of the country. Such is the case in Echegaray's talk at the opening of the 1898 course at the Ateneo de Madrid, where he proposes a theory of cultural and capital accumulation centered on the potential of human labor. This contribution to his social foundry thus involves both economic and social variables:

> De manera que una nación será grande cuando posea la más alta ciencia; cuando sea activa y trabajadora; cuando acumule grandes riquezas, y cuando alumbre á su ciencia toda la idea del deber y la idea del deber encauce todas sus energías. Y lo que digo de la nación, digo de todos sus organismos y de todos los individuos que la constituyen; que cada uno en su esfera, por modesta que pueda ser, tiene ocasión de cultivar su inteligencia, y debe cultivarla; tiene necesidad de trabajar, y debe trabajar.[12]
>
> (A nation will be great only when it possesses the highest science; when it is active and hardworking; when it accumulates extensive wealth; and when the idea of duty enlightens its science and channels its energies. And what applies to a nation also applies to all its organisms and constituencies, because in their own realm and regardless how small this domain is, each of these parts should have space to cultivate their intelligence and must do it; all of them need and must work.)

Echegaray was aware of the country's possibilities of development once education produced the necessary conditions to transform collective efforts in concrete results. His conception of society as a mechanical system regulated by physical laws and his insistence on the work ethic can be in part explained by his training as an engineer. The diligence learned during those formative years also guided his involvement in different modernizing projects, as he recalls in his memoirs in reference to the 1868 liberal revolution: "Era preciso hacer algo, trabajar siempre, demostrar con el ejemplo que la revolución no se había hecho por el gusto de hacer una revolución, sino por transformar la vieja España en una España á la moderna" (It was necessary to do something, constantly work, demonstrate by example that the revolution was not just a whim, but the aspiration of transforming the old Spain into a Spain fashioned in a modern way).[13] In his view, the revolution had represented a breakthrough in the country's modernization—it had transformed politics and education and created an unprecedented space for intellectual creativity and economic growth. By evoking this period of ideological freedom and progress-driven politics, Echegaray sets a backdrop for opposing past and present and advancing an interpretation of national reality—ideological polarization and resistance to updating "old Spain" were thus two of the main reasons that explained the country's backwardness.

Echegaray indeed saw an obstacle to modernization in some distinctive traits of the Spanish personality, such as the tendency to idealize reality or the propensity to consider problems in abstract terms. In one of his most celebrated speeches, given to the Cortes Constituyentes (Constitutional Court) in 1869, he elaborates on this issue and insists on the importance of concretizing ideas and knowledge in order to boost social transformation:

> ¿Habéis visto flotar en el cielo esas blancas neblinas, esos trasparentes tules, esas gasas de sutilísimas mallas, que ya caen en profusos pliegues en el fondo de los valles, ya se rompen en las crestas de las montañas, ya cubren pudorosamente el azul del cielo? ¿Qué son? Vapor de agua, agua diluida, agua en un estado tenuísimo de densidad, y en ese estado parece que nada son. En ese estado las neblinas del cielo son impotentes para todo; no son una fuerza: el soplo del viento las disuelve, un rayo de sol las evapora; son la idea flotante en la región del pensamiento; son la idea científica vagando en la región de las abstracciones. Es bella, es hermosa, está llena de promesas; pero como está llena de promesas toda ilusión. Mas encerrad ese vapor en las entrañas de una locomotora, dadle temperatura, dadle un organismo, dadle, por decirlo así, carne de metal, dadle palancas de acero, dadle grandes ruedas, colocadlo todo sobre dos carriles, y aquello que parecía impotente, que parecía una ilusión, se convierte en una inmensa fuerza industrial, que pasa por encima de los abismos, que rompe las entrañas de la montaña que de él se burlaba antes, y que hace estremecer el espacio con sus poderosos silbidos.[14]

> (Have you seen floating in the sky those white fogs, those transparent tulles, those subtle mesh muslins, the ones that fall in extravagant folds at the back of valleys or get broken at mountains' summits, or the ones that virtuously cover the blue sky? What are those? Water steam, diluted water, water in a state of faint density, and in this condition they seem to be nothing at all. Under that state, the sky's mists are impotent for everything; they are not strong: blowing wind dissipates them, a beam of sunlight evaporates them; they are a floating idea in the region of thinking; they are the scientific idea that wanders in the region of abstraction. It is beautiful, it is lovely, it is full of promises; but only as every other illusion is. But put that water steam in the entrails of a locomotive, give it some temperature, give it a body, give it, to put it in other words, metal flesh, give it steel levers, give it big wheels, put everything on two rails, and that which seemed impotent before, that seemed like an illusion, becomes an immense industrial force, one that is able to pass over an abyss, break open the entrails of mountains that before would defy it, and shake the space with its powerful whistling.)

Echegaray combines poetry and science to synthesize the physical principles of energy transformation, and using referents corresponding to the steam engine and the railroad operation, extrapolates from them to the realm of thought. Ideas are thus only "weak" fog, "impotent" forces until they are concretized in actions. In the same way that steam can create measurable forces once it is put inside a locomotive, concepts can generate tangible effects once they are applied to practical problems. To become useful, moreover, *ideas* need the *intervention* of people, "organisms" that in this industrial image are merely "metal flesh." Once ingenuity is channeled into concrete goals, it fuels the engines of transformation, changing potential energy into dynamism. The result of this process is an immense force able to "break the entrails of a mountain," an obstacle finally overcome thanks to the transformative capacity of industry.

Echegaray's diagnosis, according to which the national personality could be associated with a particular idealism, anticipates regenerationism's understanding of the Spanish character. For thinkers affiliated to this intellectual movement, the Romantic spirit of literary figures, such as Don Quixote, embodied not only Spaniards' inclination to resist social transformation but also their tendency to glorify the past—Ángel Ganivet's *Idearium español* (*Spain: An Interpretation*) (1896) and Miguel de Unamuno's *Vida de Don Quijote y Sancho* (The life of Don Quixote and Sancho) (1904) are good examples of this kind of assessment. This idealism, however, could be turned into concrete actions to boost social dynamism and modernization, as Echegaray illustrates with his emphasis on the capacity for work, rigor, and discipline. Industry was one of the spaces wherein this connection between material and philosophical orders became visible; hence the use of images such as the locomotive or the factory to illustrate society's ideal functioning. The dialogue established between the world of objects and the world of ideas operates in this context as a tool of calibration, a point of reference that facilitated the understanding of the complexities of modernization in a society reluctant to change.

In his talk "Aplicación de las fuerzas naturales a la industria y al comercio," Echegaray expands on this social foundry by analyzing the evolution of industry in the country—from archaic agriculture to technological manufacturing—and its importance for the national economy. The talk presents industrial activity as a body that confronts specific challenges: "Que todas las industrias en el fondo, consisten en aproximar ó separar masas finitas ó infinitamente pequeñas ó en vencer resistencias á lo largo de determinados caminos, y esto no es otra cosa más que desarrollar trabajos mecánicos" (All industries, in the end, consist in joining or separating finite or infinitely small masses, or in overcoming obstacles along the way, and that is nothing other than generating mechanical work).[15] In this view, industrialization provides society with a privileged set of possibilities or tools with which the physical and cultural obstacles of national modernization could be "overcome."

While in his address to the Cortes, Echegaray emphasized the importance of concretizing ideas, in this account of the history and problems of industry he juxtaposes the tension between theory and practice with that between tradition and progress. In order for people to effectively combine their efforts and inject dynamism into the social machinery, past and present needed to coexist:

> ¿Qué analogía hay entre un grano de trigo arrojado en el surco que ha abierto en la tierra la aguda reja de un arado, y todas las faenas agrícolas que siguen

> á esta primera, y aquel otro acto por medio del cual el minero rompe la corteza de la tierra, penetra en el interior de nuestro globo, por medio de profundísimos pozos, se extiende con galerías en uno y otro sentido, arranca el mineral del seno del filón y lo eleva á la superficie de la tierra? Entre aquel grano de trigo, arrojado al surco, y aquel pedazo de mineral extraído de lo más profundo de la montaña, ¿qué analogía hay? ¿Ni qué parecido hay entre uno y otro procedimiento? ¿Ni qué semejanza hay aún entre estos dos actos materiales y el del fundidor, que en el seno de un horno arroja el mineral, arroja el combustible, arroja el fundente, prende fuego á la masa y recoge poco después el metal líquido en moldes que de antemano tiene preparados? ¿Ni qué punto de comparación hay todavía (y permitidme que repita esto una vez más), entre todos estos actos y una locomotora que vuela sobre sus carriles, atravesando montañas, salvando abismos, uniendo de esta suerte unos y otros países, unos y otros pueblos? . . . [estos actos, estas industrias] unos y otros, y todos ellos, tienen de común una cosa; el ser trabajo humano.[16]
>
> (What analogy exists between a grain of wheat that is thrown into the furrow that the acute plowshare opens on the ground, and all the agricultural labor that follows this action, and a different operation through which the miner breaks the crust of the earth, penetrates the inside of our globe and, through deep wells, traverses galleries in one or other direction to uproot the mineral from the vein's core and bring it to the surface? What is the analogy between a grain of wheat, thrown into the plow, and a piece of mineral extracted from the depths of the mountains? What is the resemblance between these two procedures? And what similarity is there between these actions and the one undertaken by the worker in a foundry, who tosses the mineral into the furnace's core, throws in the fuel, throws in the melting element, sets the mass on fire, and collects a little later the liquid metal in molds prepared beforehand? And what element of comparison is there (and please allow me to repeat this one more time) between all these actions and a locomotive that flies on its rails, crossing mountains, passing over abysses, bringing together different countries, different nations? . . . [these actions, these industries], one and the other and all of them have in common one thing: they are human labor.)

Tracing a parallel between agricultural and mining labor, activities that consist of using the soil (plowing-excavating) and exploiting its wealth (collecting-extracting), Echegaray reconciles two incompatible temporalities (past-present) through a common factor: human labor. In this representation, tradition and progress are part of a dynamic continuum of social and material transformation—in both traditional and modern Spain, it is laborers (farmer-miner) who plant seeds (kernel-melting element) into the core (soil-furnace); there, different natural forces operate (gestation-heat) to produce fruits (edibles-liquid metal). This image is completed with a description of the locomotive's speed, a materialization of the use of science: trains, fast as birds, were able to pass over different natural obstacles (mountains, abysses) and gather people in a larger and stronger imagined community.

Yet the country's economy could not be supported only by labor or capacity for work; the national engine also needed fuel to function properly. In that regard, natural resources were deemed essential for industry and commerce. For Echegaray one of the nation's main sources of wealth was its fortuitous geographical location. Since the energy in minerals such as coal or iron came directly from the sun, it was natural that Spain, whose advantageous location within the continent gave it matchless exposure to the sun's light, had important reserves of energy and the potential for turning it into wealth:

> Vosotros veis que la locomotora marcha sobre sus carriles, arrastrando un largo tren. ¿Cuál creéis que es la fuerza que mueve esa locomotora? . . . Esta fuerza, repito, no es más que la fuerza solar de hace muchos siglos. El sol que ardió hace cien siglos ó mil siglos há ¡quién calcula estas cifras! ese sol es el que hoy arrastra todas las locomotoras del mundo sobre sus carriles. . . . Trasformado después este trabajo en calor, y el calor pasando al vapor de agua, arrastrará la locomotora por sus carriles, llevando consigo todos los productos de la industria de un pueblo á otro pueblo, de una á otra región. Hé aquí, como os decía al principio, de qué manera el sol de hace muchos siglos es el que hoy trabaja en nuestros caminos de hierro.[17]
>
> (You can see the locomotive running on its rails, pulling behind it a long train. Where do you think the force that moves this locomotive resides? . . . That force, I repeat, is no other thing than the sun's energy from many centuries ago. The sun that shone one hundred or a thousand centuries ago—who calculates these things! That sun is the one that today moves on their rails all the locomotives of the world. . . . Transforming this energy into heat, and then using it to produce water steam, that sun's power will push the locomotive on its rails, carrying all the products of industry from one town to another, from one region to the other. That is, as I mentioned earlier, the way in which the sun from many centuries ago is working today in our railroads.)

Through multiple processes of energy transformation, trains were not only able to move and relocate assets but also to function as vehicles for the circulation of ideas. Echegaray's depiction of this activity, however, commends a centralized vision of power: all energy emanates from one center—the sun—to later be distributed and reconverted elsewhere. Analogously, ideas and projects for the country originated in the capital, where they were discussed and enforced to produce concrete changes in other parts of the peninsula.

It was true that Spain was behind other European countries in terms of scientific and technological development, but its geographical superiority somehow compensated for the lack of concrete actions to generate social transformation. If anything, this advantage was the source of a youthful and generous national spirit, one that, given the proper conditions, could elevate the country to the level of other industrial powers:

> Sus minas, las de los ingleses, son subterráneas, son lóbregas, son oscuras; nuestras minas serían la bóveda azul del espacio: los ingleses tendrían, digámoslo así, un sol fiambre [Grandes risas], un sol viejo, y nosotros tendríamos un sol fresco y palpitante, porque nosotros en España, ó no hacemos las cosas (y generalmente no las hacemos), ó cuando las hacemos, las hacemos de una manera expléndida [*sic*] y generosa [Nutridos aplausos].[18]
>
> (Their mines, the British ones, are underground, unilluminated, dark; in contrast, our mines are like the blue celestial sky: the British have, if we put it in that way, a chilly sun [loud laughing], an old sun, and we [Spaniards] have a fresh, vibrating one, because in Spain when we do things—and generally we don't do them—we do them in a splendid and generous way [loud round of applause].)

Echegaray's proposal was an invitation to make the most of the country's exceptional circumstances to create the conditions that would later facilitate scientific and technological development. By reducing human and material costs in the advancement of the economy, this strategy resonated with some of the suggestions made by free-trade schools of economic thought, perspectives that the Circle of Commercial Alliances of Madrid (where this talk was delivered) advocated and was promoting through Echegaray's cycle of lectures.[19]

Some aspects of scientific and technological materialism, however, could not be avoided in the implementation of economic and social reforms. In the collective imagination of the period, modernization was a menace to tradition, particularly to religious spirituality—science questioned the existence of God and diminished humanity by reducing perception and consciousness to mere chemical reactions that took place in the brain.[20] In spite of being a liberal thinker, Echegaray openly rejected this kind of materialism: "La opinión que combato es fórmula menguada y repugnante del más embrutecedor materialismo; tal doctrina desconoce por completo las más nobles, las más puras, las más elevadas facultades del hombre; le mutila torpemente, y sin piedad, le reduce a inmunda bestia" (The opinion I reject is a diminished and repugnant formula of the most stupefying materialism; such a doctrine lacks the most noble and pure human attributes; it clumsily mutilates man, and without any commiseration reduces him to a disgusting beast). And then he adds, "Es el hombre realidad, más intensa, y más rica, y más elevada, que el mundo de la materia y de los sentidos" (Man is a more intense, rich and elevated reality than the material or sensorial world).[21] Following Krausist precepts, Echegaray attempted an integration of materiality and the conservative principles of Catholicism while defending the progressive nature of his political agenda: "Ya he dicho que en estas materias yo era y soy católico-ortodoxo y viejo creyente en las leyes eternas de la Economía" (I have already said that in these matters I was and still am an orthodox Catholic and an old believer in the eternal laws of Economy).[22] Only in a scenario wherein tradition and progress could harmoniously coexist, modernization would finally be seen as a positive force and a necessary step in the country's development.

Echegaray's ambition to find a balance between spiritual and material forces was not limited to economic or political ideas; it also configured his work as a writer. Reflecting on the tension between creativity and interpretation, in his inaugural speech as a member of the Royal Spanish Academy he complemented his social foundry by advancing the idea of an aesthetic and literary *science*—a tool that could be used to detect and counteract attitudes of ideological intolerance:

> Así como importa mucho para la marcha ordenada de la política, sobre todo en épocas de transición, que exista una legalidad común, no menos importa en el campo artístico y literario otra especie de legalidad común, dentro de la cual vivan y se desarrollen pacíficamente todas las escuelas y todas las energías, sin anatemas ni excomuniones desde arriba, sin odios ni enemigos desde abajo.[23]
>
> (In the same way that it is important for the organized advancement of politics—mostly in times of transition—to have a common legality in place,[24] not less important is having in the artistic and literary fields another kind of common legality, one in which all schools and all energies harmonically evolve without anathemas and excommunications from above or without hate and enmity from below.)

If the exercise of literary criticism needed to be reformulated, it had to resemble a scientific discipline in which an aesthetic phenomenon could be explained in universal terms—experimentation, methodology, and classification, therefore, had to remain essential to its practice. The scientific method, with its natural objectivism, thus extended its possibilities to the subjective appreciation of artistic creation. Literary critics' main task, in turn, consisted in defining the rules governing this science. With this proposal, Echegaray hoped to incorporate the *aesthetic* aspects of modernization into his assessment of the country.

Cultural dynamism could be described in terms similar to that of industrial transformation, and it was ruled by comparable principles; consequently, an unbalanced distribution of cultural resources could be damaging to society. As noted earlier, thermodynamics offered scientific arguments to justify disparity—an unequal distribution of wealth, for example, was indispensable for the circulation of labor and the transformation of effort into tangible goods. Following these ideas, in his social foundry Echegaray contends that one of the central priorities of modernization was to produce social balance while keeping different and distinct class roles: "Las facultades humanas son múltiples, pero deben ser armónicas; y todo desequilibrio en el individuo ó en la sociedad, como en el arte, es causa de decadencia, y al fin es destrucción y ruina" (Human potential is diverse, but it must be harmonic; all individual or collective instability, as it also happens with the arts, causes decay, and in the end implies destruction and ruin).[25] One of the main causes for the country's problematic development was the instability of the present. In order to successfully translate this social strain into the literary world, Echegaray establishes a contrast between the aesthetics of the classics and that of naturalism, thus showing two different sensitivities that coexisted under the conditions of modernization.

For Echegaray both tradition and progress deserved respect and consideration; it was important to revere the glories of the past, but it was also essential to recognize the potential of the present:

> Pasad del mundo clásico, puro y correcto como las líneas armoniosas del templo griego, al realismo brutal de ciertos escritores, ó mejor dicho, al materialismo moderno; pasad, digo, de aquellas limpias y trasparentes formas literarias, en que hasta los pastoriles cariños del pastor Goridón y del hermoso mancebo Alexis, delicias de su dueño, se velaban con el ritmo del exámetro [*sic*], á los bestiales amores de *Jerminal* [*sic*], . . . y tendréis otra literatura . . . digna de censura por sus extravíos, pero que no sería justo rechazar por completo, ni condenar para siempre; que, después de todo, . . . entre el fango de las negras galerías, aun pudiera entresacarse del polvo negro del cok más de un pedazo de brillante, pues al fin y al cabo, carbón cristalizado, nos dice la ciencia, que es el maravilloso cristal de los metálicos y azulados reflejos. . . . Y ya tenemos la primera oposición: el mundo clásico y el mundo moderno. Pues yo, en nombre de esa legalidad común de que os hablaba, pido justicia para ambos: respeto, admiración y estudio para aquél; respeto, esperanza y alientos para éste.[26]

> (Move from the classic world, pure and accurate as the harmonious lines of a Greek temple, to the brutal realism of some writers, or, to put it in another way, to modern materialism; move, I say, from those clean and transparent literary forms in which even the rural affections between Goridon, the Shepard, and young and beautiful Alexis, favorite of his master, were carefully developed using hexametric rhyme, to the beastly passions of *Germinal*, . . . and you will have a different literature, . . . one deserving censorship because of its deviations, but one that it would not be just to entirely reject or censure; because after all, . . . within the mud of the black galleries, there is always the chance to recover more than one diamond from the black dust of coal; in any case, as science teaches us, the marvelous crystal with metallic and blue glares is nothing but a piece of crystalized coal. . . . And we have the first opposition:

classic and modern worlds. Well, in the name of that common legality I was talking about earlier, I demand justice for both: respect, admiration, and analysis for the first; respect, hope, and encouragement for the second.)

Echegaray's comparison here transcends the literary realm and points to a social disposition wherein the past is associated with positive ideas of purity and perfection while modernization is depicted as a negative, contaminating force casting a shadow on tradition. In this example, mining, in spite of representing darkness, filthiness, and extreme labor conditions, hides the wealth that can fuel the country's economy. Similarly, naturalism's sordid particularities encompass new forms of beauty that surpass the unfavorable aspects of its characteristic materialism; if aesthetic experiences could be adjusted to this modern sensitivity, there also had to be conceptual mechanisms to harmonize tradition and progress in other areas of society.

Consonant with this prospect, Echegaray criticizes certain approaches to literature that, by using the classics as aesthetic referents, tended to misinterpret contemporary reality. By suggesting a renewal of the critical practice, his talk at the Academy also acknowledged the need to push the country into the dynamism of European modernization. Echegaray thus concluded: "He procurado demostrar, finalmente, que en estas épocas de transformaciones es cuando más amplia, más generosa, menos terca debe ser la crítica, para no exponerse á grandes errores, contrariando sistemáticamente el desarrollo de gérmenes fecundos" (I tried to show, in conclusion, that it is in these times of transformation when criticism needs to be more flexible, more generous, and less obstinate in order to prevent big mistakes, systematically opposing the development of fertile seeds).[27] The sort of *industrial* hermeneutics that stemmed from integrating scientific, technological, and cultural preoccupations into society's interpretation yielded a better understanding of the principles governing the modern world. This perspective not only revealed the adaptability of the national character but also the possibilities of creating contiguities between past and present in the context of modernization.

In spite of Echegaray's clear effort to underscore the importance of science and technology in the economic progress of the country, his social foundry does not delve into the country's economy and the possible models for improving it. Although ideas on productivity and the circulation of wealth and labor are suggested in his analyses, he does not offer a systematic dissection of the advantages and disadvantages of adopting free-trade strategies or embracing protectionist measures. This work, nonetheless, would be completed by other thinkers whose particular position within the material field provided them with the tools to assess the country's resources and evaluate modernization's main economic obstacles. In what follows I review the ideas of two of these scientists to show how their social foundries propose an insightful diagnosis of Spain's fiscal and monetary afflictions.

Pere Estasen's Energetic Protectionism

Pere Estasen i Cortada's programmatic proposal encompasses two opposite economic approaches—his goal was to implement a mixed economy in the country wherein free trade and protectionism could coexist to serve domestic and international commercial demands. This model was enlightened by his understanding of Darwin's theory of biological evolution and of Walter Bagehot's economic views, and reflects in part Estasen's own nationalistic views with regard to Catalonia—as well as his profound understanding of Spain's strengths in terms of both human capital and natural resources. In his capacity as a lawyer, Estasen had privileged access to the public sphere in his native Barcelona, a condition that he capitalized on to consolidate an audience for his work and to promote the scientific positivism and economic ideas of philosophers such as Herbert Spencer, Friedrich List, and Henry Carey. Through his activity as a researcher, Estasen identified that one of the critical problems of the Spanish economy was the accumulation of capital in a few sectors of society. For him, this unequal distribution of wealth, along with investors' short-term vision, had pushed capitalists into taking unnecessary risks in the stock market and, in many cases, getting involved in speculative activities that only contributed to the country's financial instability.[28] Estasen believed that Spain needed a solid program of public investment that could be channeled through state institutions, such as the Ministerio de Fomento (Ministry of Development); this public policy would guarantee funding diversification and wealth redistribution, thus promoting modernization through the consolidation of industry and the improvement of social conditions.[29]

As pointed out earlier, capital accumulation, allocation of resources, and labor power were economic markers that resonated with some of the physical theories that circulated and became popular during this period.[30] Against this background, the configuration of Estasen's social foundry offers relevant information about the ideological negotiations that shaped the *fin-de-siglo* economy. In what follows, I analyze his influential 1880 essay *La protección y el libre cambio: consideraciones generales sobre la organización económica de las nacionalidades y la libertad de comercio* (Protection and free trade: general considerations on the economic organization of nations and deregulated commerce) to show how industrial and scientific referents inform and shape his assessment of the country's economic challenges vis-à-vis his ideas on identity in the context of a plurinational state such as Spain.

Estasen's economic thinking was supported by his understanding of the principles of evolutionary biology. Following Darwin's ideas on biological competition, he believed in the existence of a constant struggle for survival that extended to all levels of society; within the economy this mechanism was regulated by the forces of the market. In the case of Spain, the implementation of a liberal economy had led to political fragmentation, thus unbalancing the distribution of social forces and complicating the possibilities of economic cooperation:

> Mas como en la naturaleza y en la sociedad todo tiene su regulador y su equivalente, ya estamos tocando las consecuencias de este exagerado predominio, pues si el espíritu mercantil se ha impuesto y dominado, ahora se encuentra cuarteado y dividido, consecuencia de su excesiva preponderancia. La supremacía de algunos elementos económicos y su lucha intestina han provocado la crisis. Recordemos un fenómeno, cuyas leyes esplican [*sic*] las ciencias naturales. Varias especies de animales habitan una circunscripción determinada; dispútanse el alimento, la habitación, en una palabra, las condiciones favorables del medio ambiente en que viven.[31]
>
> (Since in nature, as in society, everything has a regulator and an equivalent, we are already seeing the consequences of this exaggerated prevalence [of economic liberalism], because even when the commercial spirit has been imposed and dominates, now it is cracked and divided as the result of its own preponderance. The preponderance of some economic elements and their internal struggle have provoked a crisis. Let us remember here a natural phenomenon whose rules are explained by the natural sciences. Different animal species share a particular habitat, they compete for food and shelter; in a single word, they fight for access to favorable conditions in their habitat.)

It was precisely the sustained competition to adopt a particular economic approach that complicated the social coalescence necessary to advance modernization.

Biology, nonetheless, was not the only scientific referent in Estasen's economic analysis. Capital, money, and value, for example, could be seen as components in a complex compound wherein the interaction of molecules and atomic particles reflected various aspects of social behavior. If scientists were able to understand, model, and predict the interaction of these elements by appropriating laws, concepts, and theories related to physics, chemistry, or mathematics, economists could do the same in society:

> De la misma manera que los átomos se atraen en el mundo físico, así parece que se atrae el dinero en el mundo económico y social, y el individuo se halla imposibilitado de hacer algo, como no tenga un capital, así sea este fijo, así circulante, pero siempre un capital; de la misma manera que no puede desenvolverse una série orgánica, ni se ofrece fenómeno de vida, de nutrición, de relación, etc., si no hay una série de compuestos de carbono ó una simple célula.[32]
>
> (In the same way atoms attract each other in the physical world, so apparently does money in the economic and social realms; and an individual finds himself

> unable to take action if he does not have capital, either fixed or variable, but all the same capital, in the same way that an organic compound cannot evolve, or possibilities for life, nutrition, relationships, and so forth do not exist if there is no carbon compound or a simple cell.)

Capital accumulation operates in this image as potential energy, so does the idea of labor power. By analogy with the principle of energy conservation, the circulation of money could only be guaranteed insofar as there were reserves of capital, whether these were in the form of natural resources, labor, or material wealth.

In the essay a great degree of attention is paid to connections between national wealth and human capital, both of which were expressions of two factors: labor power (Arbeitskraft) and abundance of natural resources. One of the main obstacles to capitalizing on this potential energy was accessibility. Industrial modernization was key to improving the circulation of people and goods through the development of more efficient means of transportation and the expansion of other communication networks. The particular disposition of these networks, however, had a profound impact on the way people and products were distributed, changing costs of production according to factors such as availability, shipment times, and so forth:

> La sávia nutritiva de las naciones se distribuye de una manera desigual por la organización especial que preside al aparato encargado de esta función ó sea la red de los ferro-carriles; y así sucede, que mientras la vida se concentra en una ciudad y la sangre se congestiona por decirlo así en un órgano, el resto del cuerpo está falto de la sávia indispensable. El aparato circulatorio del organismo económico no funciona bien y la salud y la vida se hallan en constante peligro. Los ferro-carriles han puesto en comunicación los grandes centros de población de todas las naciones, al mismo tiempo que han alejado más y más la producción de la aldea, de la producción de la ciudad, los productos del villorrio de los productos de la casa de campo y de la finca rústica aislada. Cerca de la línea de un ferro-carril, el producto vale mucho porque fácilmente se transporta á un punto de consumo. La finca próxima á la ciudad donde la vía empieza ó termina, aumenta su valor de día en día. En cambio, el excelente producto que se elabora en las poblaciones intermedias no encuentra tan fácil su salida.[33]

> (Nations' nutritious sap is unevenly distributed due to the particular organization of the instrument that serves this function, namely the railroad network; and then it happens that, while people concentrate in cities, the circulatory system converges, so to speak, in a single organ, the rest of the body has limited access to the indispensable sap. As a consequence, the economic organism does not work properly, and its health and life are under constant

threat. Railroads have connected the population centers of most nations, but, at the same time, they have more and more separated provincial from urban production, city products from country houses' and rustic farms' goods. In the proximity of the railroad line products have a higher value because they can be easily transported to the consumer. Conversely, excellent products of intermediate cities cannot be so easily commercialized.)

Yet industrialization's contradictory effects on the economy, illustrated here in the positive and negative aspects of the railroad, reveal another problem: the separation between cities, where modernization had concentrated, and the countryside. Although the railroad connected center and periphery in an unprecedented manner, this change in the way in which capital and goods circulated also affected the economy. Despite the efforts of urban planners and engineers, multiple natural and human resources remained allocated outside the limits of the transportation network. Estasen was a staunch critic of this incidental effect of material progress; he recognized the importance of the railroad for the country's welfare, advocating for investment to continue expanding its reach, and was convinced that a better integration of rural and urban spaces would facilitate the exploitation of resources and improve the economy.

But the topology of transportation networks was not the only factor that affected wealth distribution and labor power in Spain. The consolidation of the national economy was also contingent on the interference of foreign capitals. Nationalization of resources and shielding of local markets were for that reason essential components of Estasen's defense of protectionism—they constituted a mechanism for bringing together private and public investment in the country's interest. One of the most damaging elements to the local economy was foreign companies' control over the natural resources, a practice that redounded to the impoverishment of rural areas and the deterioration of working conditions:

> Económicamente hablando, nuestra patria está desorganizada y solo puede reorganizarla un régimen protector.... Ya hemos visto anteriormente que los grandes capitales, las grandes empresas, los grandes negocios de España, están en manos de extranjeros. Mas adelante veremos como uno de los principales obstáculos á nuestra regeneracion económica, son las grandes compañías de ferro-carriles, en cuyos Consejos de Administracion se encuentran los jefes de la política y los hombres mas influyentes del país, espléndidamente retribuidos.[34]

> (Economically speaking, our country is disorganized and only a protectionist regime can change that.... We have already seen earlier how large capitals, big companies, big businesses in Spain are in foreign hands. Later we will see how one of the main obstacles for our economic regeneration is the large railroad companies, whose boards of trustees have handsomely compensated political leaders and influential men of the country.)

Estasen's social foundry points to disorganization as one of the most serious symptoms of the national illness. While the second law of thermodynamics establishes that with time a system irreversibly loses its capacity for producing work, the suggestion here seems to disregard this principle, aiming at changing the direction of time so that from an already entropic state, Spain's economy could regenerate. Following this logic, it was not surprising that conservatives looked at tradition to find in the past the necessary elements for healing the country. In this vision, traditional Spain was organized and *clean* on many levels: racial, linguistic, religious; rejecting foreign influences—be they economic or cultural—was then a natural response of conservatism to social change. A protectionist approach to the economy offered Spain the needed strengthening of its local forces and the consolidation of its national pride.

By reaffirming the strong links that existed between financial interests and political and cultural agendas, Estasen shows how the main traits of national identity both shape and are shaped by the economy: "La manera de que España sea una é indivisible, es unir el vínculo económico al vínculo político, de lengua, de costumbres, de religion y tradiciones, que en mayor ó menor escala ya existe" (The only way in which Spain can be an indivisible unity is by creating a link, that to a large extent already exists, between the economy and politics, language, customs, religion, and traditions).[35] If one considers the suggested connections between material modernization and national culture, science and education surface as determining factors in the country's economic prosperity. For Estasen, as for many other thinkers studied in this book, national development relied on the existence of a robust system of science education. Hence the certainty that Spain would not be able to progress "while science was poor and ignorance rich."[36] A protectionist model could facilitate the conditions of competition and capital accumulation necessary for the development of higher education to the levels demanded by modernization. This experiment had already been tried in other European countries and in the United States, as Estasen points out in his essay, where economic recovery attested to the advantages of protectionism.[37] The regeneration or strengthening of these foreign economies had been in part the result of the local intellect and the political institutions already in place. If Spain wanted to emulate these processes, it would have to work on two fronts simultaneously: reconciling a collective view of the future wherein the traits of identity could be potentialized to facilitate modernization, and creating an adequate institutional infrastructure for the development of science and technology.

Given its current conditions, it was evident to Estasen that the country needed to work intensively on improving its scientific capital. A solid base on this front would improve productivity on multiple levels, minimizing the need for international intervention and maximizing the use of local human and natural capitals: "Las naciones más científicas, más civilizadas, más adelantadas, son las que dominan y las que se encuentran en mejores condiciones de procurarse las riquezas que han de proporcionarles medios para adelantar en el arte, en la ciencia, en la comodidad de la vida, en la perfección del trabajo, en la fuerza de las armas y en las condiciones etnológicas de sus habitantes" (The most advanced, civilized, scientific nations are the ones that dominate and have better mechanisms to produce the means required for promoting their arts, science, comfort of life, quality of work, military force, and ethnic conditions of their inhabitants].[38] As a fin-de-siglo thinker, Estasen believed in the influence of geographical and racial factors as defining traits of national identity. As historian Joshua Goode demonstrates, the idea that religious, social, and cultural practices were attached to specific ethnic conditions owed its popularity to the consolidation of anthropology as a scientific discipline.[39] From this perspective, the country's wealth was, to a great extent, rooted in its diversity (natural and human); improving the means to study and assess this heterogeneity was thus essential to boosting the economy and modernizing the nation.

The state's role in this process of linking economic development to national identity was heightened by the creation of local institutions whose mission consisted of facilitating progress while protecting tradition and history from external influences:

> El Estado debe proteger ante todo el elemento histórico, las instituciones en que se encarna la tradición, los intereses morales, limitándose aquí la protección á prestar recursos y á procurar la defensa contra todo ataque externo é interno; debe fomentar el elemento científico, así personal como material, pues vale más á una nación un génio que se llame Edison ó Newton, Fortuny ó Cellini, que todas las fábricas reunidas.[40]
>
> (The state must protect, primarily, its historical character, the institutions that embody tradition, the moral interest; and this protection is understood here as the creation of mechanisms that provide resources and ensure the defense against any external or internal attack. [The state] must promote the scientific spirit, at personal or material levels, since for a nation an individual genius like Edison or Newton, Fortuny, or Cellini is more valuable than all factories together.)

Since the human element was more important than any material force, the idea of education as the engine of progress acquires a new meaning. Because it provided the necessary conditions to develop intellectual capital and consolidate science and technology as national values, education was the social machinery's ideal fuel. The mission of the state within this framework was therefore to protect this capital from foreign interference. Estasen's protectionism was thus not only conservative in economic terms but also in its approach to society's cultural configuration—nations constituted complex organisms in which the state, through a centralized government, operated as a regulatory entity able to police and preserve the wealth of the country.

Estasen's social foundry was supported on a teleology derived from evolutionary theories—the same logic that underpins the first and second laws of thermodynamics, scientifically justifying the constant transformation of society and its progress toward a final stage. Furthermore, his reading of society defended the need for a rigid structure of classes wherein the aristocracy not only persisted but also held a privileged position that neither the forces of capital nor a revolution could overthrow:

> En la naturaleza y en la sociedad nada se improvisa; y lo que por evolución lenta y gradual se desarrolla, es robusto y sólido; de ahí que la política positiva y científica recomiende la evolución social y condene las revoluciones; de ahí que abogue por el sostenimiento de las clases sociales, la estabilidad y firmeza de las capas que constituyen la sociedad, la categoría de profesiones, la dignidad de los cargos, el respeto á la tradición y el gran prestigio á las clases que la representan; en una palabra, la necesidad de una aristocracia, con gran prestigio social, prestigio que no se sostiene cuando los bruscos movimientos del cuerpo social desequilibran la natural posición de las clases, y las familias de ilustre abolengo quedan en pocos años reducidas á la nada y han de descender á las últimas clases para poder vivir, mientras que hombres audaces, ó con suerte, de mozos de cordel ó revendedores, en pocos años han pasado á ser hombres millonarios y por consiguiente hombres de prestigio é influencia.[41]

> (In society as in nature nothing is improvised; and what develops through a gradual and slow evolution is robust and solid; hence positivistic and scientific politics recommend social evolution while criticizing revolution; for that reason, they support the preservation of social classes, the stability and firmness of the layers that form society, the categorization of professions, the special dignity of hierarchical positions, the respect for tradition and the standing of the social classes that represent it; in a word, the need for an aristocracy with substantial social status, a prestige that is not possible when abrupt movements of the social body destabilize the natural configurations of class and families of distinguished lineage get reduced in a few years to nothing, forced to descend to lower classes to survive, while daring men, or just lucky ones, dock workers or retailers, in a few years have become millionaires and therefore prestigious and influential men.)

Social disorganization, criticized from the beginning of the essay, is to a great extent the result of industrialization and its new systems of capital accumulation and circulation of wealth. Fear of losing the power structures that guaranteed privileges for the higher classes is one of the reasons Estasen praises individual talent over collective development as a token of modernization. While these social schemes, closer to a feudal system than to an industrial economy, were obsolete, they stood for a lost greatness that many authors yearned to restore. By looking at the past, Estasen's social foundry aimed at violating the second law of thermodynamics; a contradictory approach that was at odds with his own evolutionary perspective on society.

While in Estasen's view factors such as race and social class justified the implementation of a particular economic system, for other thinkers it was geography and natural resources that needed to be considered in diagnoses of the national organism. Whether the country was sick or not, its potential for improving resided in the wealth of its lands, its geographical position, its climate, and the natural disposition of its peoples. As I show in the next section, such was the line of analysis of geologist Lucas Mallada and the basis for his social foundry.

Lucas Mallada's Geology of the National Problems

Lucas Mallada's groundbreaking essay *Los males de la patria y la futura revolución española* (The ills of the country and the future Spanish revolution) (1890) is both the work of a celebrated scientist and an accomplished literary piece. Mallada, a geologist, engineer, and prolific writer, devoted his life to research and teaching—he held a position as professor of paleontology in the Escuela Técnica Superior de Ingenieros de Minas y Energía (Advanced Technical School of Mining and Energy Engineering). His important contribution to the aforementioned *Mapa geológico de España* led him to publish in the last decade of the century a series of reports about the national geography, numerous scientific articles in his field of research, and a pioneering catalog of the fossil species of Spain. Mallada also wrote several essays; in addition to *Los males de la patria*, it is worth noting here his epistolary compilation *Cartas aragonesas dedicadas a S.M. el Rey Alfonso XIII* (Aragonese letters written for H.M. the king Alfonso XIII) (1905). In both texts, Mallada makes a painstaking exposition of what he considered the essential issues facing the country.[42] Even though he was not directly involved in politics (Mallada declined nominations for the positions of ministry in the central government and mayor of Madrid), his ideas impacted many politicians with whom he shared pessimistic views of the country's institutional decay at the turn of the twentieth century, as well as progressivist perspectives on the future of education.

As in the case of other writers studied in this book, Mallada's scientific knowledge, field work, and technical expertise shaped his approach to the analysis of society. His social foundry is in that sense not strictly the result of having incorporated notions, images, and objects of the industrial world into his narrative universe, but of directly applying his scientific understanding of reality to social problems; it responded to what geographer Steven Driever describes as the natural convergence of nineteenth-century scientism and an incipient anthropological insight: "Mallada's generation of scientists was still educated only in the broadest sense. . . . Like Auguste Comte, the founder of positivism, they considered social and ethical issues to offer serious possibilities for investigation by the scientist, more serious than some 'static' fields of study within the natural sciences."[43] As a consequence, Mallada advanced a particular reading of the complexities of industrialization, their relation to modernization, and their importance for the construction of a modern national identity. In what follows, I analyze *Los males de la patria* as a complex diagnosis of modernization in Spain; to complement that reading, I also allude to Mallada's inaugural address at the Royal Academy of Sciences in 1897, a talk in which he further develops his social foundry by presenting a contrasting overview of the history of geology in the country as yet another example of the difficulties in building a national scientific tradition.

The pessimistic tone of a title such as "Los males de la patria" predisposes the reader to understand the nation as problematic and ailing. Based on rigorous scientific research, to prove its arguments the essay deploys a sophisticated conceptual apparatus that relied on the use of advanced statistical tools. Yet the study's scope transcends the technical realm to explore specific social and cultural tensions and present a diagnosis of the country's different afflictions. Because of its versatility, interdisciplinarity, and literary prose, the essay reached a wide range of audiences and gained notoriety within well-stablished intellectual circles around the country. Joaquín Costa, for example, head editor of the *Boletín de la Institución Libre de Enseñanza* (Bulletin of the Independent Institute of Education) at the time of the text's publication, promoted the dissemination of its findings and their discussion in open debates that were held at the Ateneo de Madrid. As an ideologist of regenerationism,[44] Costa enthusiastically endorsed the essay, highlighting its poignant critique of Spain's alleged moral superiority—an assumption that from the regenerationists' perspective had prevented a better understanding of the causes for the national decay. Costa even advocated the publication of the full essay in the pages of the *Boletín*, a decision that gave unexpected popularity to both Mallada's text and the journal.[45]

The text's ample diffusion and its controversial ideas on the causes of the national problems, however, drove Mallada to take part in heated public debates, in which many of his opponents accused him of projecting a skewed view of the country and of having based his work on speculation. These reactions, according to Driever, were far more negative than the author or Costa had anticipated; for example, one of Mallada's colleagues in the Comisión del Mapa Geológico, engineer Federico de Botella, gave a talk in 1882 at the Geographical Society entitled "De cómo nuestro suelo no es tan pobre como se quiere decir" (On how our soil is not as poor as it is said). Botella not only refuted Mallada's argument point by point, but he also slammed his methodology and results, which he considered biased and unnecessarily pessimistic. Critical of the essay in similar terms as Botella were geologists Martín Ferreiro and Francisco Coello, who, surprisingly, were also part of Mallada's team in the Comisión del Mapa Geológico. For them, the essay's data analysis was not rigorous enough, and the results lacked the necessary objectivity of a scientific study.[46]

There are multiple connections, as well as clear differences, between Mallada's and Echegaray's approaches to the problems of Spain. Whereas the latter, as pointed out earlier, had an optimistic view of the country's possibilities in the context of modernization, framing his discussion within notions of potentiality and labor power, Mallada insisted on tackling the nation's difficulties through a radical reformulation of the national character. He was aware of the subtle but consistent efforts of a small scientific community whose achievements could be propelled by a better understanding of identity. A gradual transition of the country from outdated social, political, and economic orders into structures compatible with industrial reality was thus essential for incorporating science and technology as active components of national transformation. Mallada's view of industrialization and identity as interconnected processes resonates with Gellner's ideas on the origin of nationalism. For Gellner, "industrialization engenders a mobile and culturally homogeneous society, which consequently has egalitarian expectations and aspirations."[47] Scientific and technological advancement were thus essential not only for progress in general but also for the emergence of a sense of belonging to the shared symbolic orders of the nation. Within the regenerationist ideology, as economic historian José Luis Ramos Gorostiza explains, identity was understood in a Romantic way, namely as a collection of shared traits that formed a sort of *national personality*.[48] Regenerationists looked back to German Romanticism and recovered the concept of nation as a dual entity consisting of body and spirit. The importance of psychology within this perspective lay precisely in the notion that countries possess a collective mind and an organic memory.[49] In Mallada's view, part of the national *psychological* disarray originated in the extended belief of the country's exceptionality in Europe, a position held, among others, on account of its geographic location and the natural wealth of its soil.

Los males de la patria contests the idea of Spain's singularity from the very beginning. For Mallada it was this vision that had yielded many misconceptions about the country's wealth, economy, and culture: "Tan arraigada se halla en España la creencia de que vivimos en un país muy rico y de muchos recursos naturales, que no sin cierto encogimiento nos permitimos decir algo en contrario, pidiendo ante todo perdón a los que desde el comienzo nos tachan de pesimistas" (The idea that we live in a very wealthy country, full of natural resources, is so deeply rooted within us that it is not without feeling some shame that we dare to say something different, mainly offering apologies to the ones that from the beginning accuse us of being pessimistic).[50] The text's central argument transcends the moral and psychological evaluations that characterized other regenerationist analyses, to instead delve into the scientific study of geography and produce a precise assessment of the country's actual wealth. From this perspective, the low productivity of the agrarian sector, for example, rather than being attributed to factors such as the supposed congenital laziness of the Spanish individual or his alleged adventurous spirit, is presented as a consequence of the natural constitution of the soil. Due to climatic or geological circumstances, arable land in the center of the peninsula remained unproductive during most of the year. Assuming the perspective of a foreigner visiting these regions, in the text Mallada wonders, "¿Qué idea queréis que se forme de la riqueza de nuestro país el extranjero que circule por casi todas las vías férreas? Si penetra en España por Irún, en cuanto pasa el Ebro, a sus ojos se presenta Castilla la Vieja, tan seca y tan desarbolada, que más fundado hallará el nombre de vieja por lo decrépita y poco florida" (What idea do you want visitors to have about our country's wealth when they travel on any of our railroad lines? If the traveler enters Spain through Irún, as soon as he crosses the Ebro he would see Castilla la Vieja, a territory so dry and treeless that he would think that the adjective "old" refers to its decrepit and colorless condition).[51] In fact, as he notes, this natural precariousness had somehow defined the personality of this region's populace, emphasizing traits such as distrust or rudeness:

> En las comarcas escasas o privadas de arbolado, las cualidades morales de sus pobladores son menos apreciables que las de otras cuya existencia corre venturosa entre una rica vegetación. En éstos veréis muchas señales de cultura; en aquéllos, la sequedad del suelo engendró la sequedad del espíritu y produjo la rudeza y los feroces instintos. No estimuléis su inteligencia embotada; no os inquietéis por cultivar su educación. Rechazan cuanto tienda a mejorar sus condiciones sociales y se consideran dichosos en su abandono y en su estado próximo al idiotismo. Mas si por compasión o por interés nacional os avergüenzan tales compatriotas, dadles agua a todo trance, cambiad el aspecto de su país, y habréis hecho una nueva conquista en provecho de la civilización.[52]
>
> (In villages with few or no trees, settlers' moral qualities are less marked than those of people whose existence luckily takes place amid rich vegetation. In the latter case you will see many signs of culture; in the former, the dryness of soil engenders a dryness of the spirit, producing rudeness and feral instincts. Do not bother stimulating their muddled minds; do not worry about cultivating their education. They reject everything that tends to improve their social conditions, and they consider themselves happy at their negligent and almost-idiotic condition. But if you feel moved by compassion, or if in the interest of your nation you happen to feel embarrassed by these compatriots, first of all provide them with water, transform the aspect of their town, and you will have then made a new conquest for the benefit of civilization.)

The reductionist determinism of this assessment was extremely controversial not only for its materialistic tone, probably inspired by readings of German geographer Friedrich Ratzel,[53] but also for suggesting that Spain was not necessarily a civilized nation. Mallada connects here geography and identity to suggest a transformation of the soil that would help the economy and, at the same time, civilize the national personality. This refashioning could only be understood as part of a scientific endeavor in which geology played a pivotal role. Whereas Echegaray's social foundry demanded the transformation of the nation's potential energies into palpable work, Mallada's assessment recommended a renovation of the soil to ameliorate the limitations that the national character had imposed on the country's economic possibilities.

Mallada's emphasis on the national essence as an expression of the country's geological configuration, however, relies for the most part on the exceptionalism he wanted to criticize. Strongly influenced by authors such as German philosopher Johann Fichte, Mallada's analysis deploys a concept of race rooted in the same geographical determinism that structured most of his argument. The popular belief according to which Germanic races were superior to Latin races, for example, informs some of the central ideas of the essay. This is not surprising if one considers the proliferation at the time of pseudoscientific ideas of racial classification, the emergence of social Darwinism, or the regenerationists' interest in the Europeanization of Spain. Following some of these ideas, Mallada blames the country's decline on Spaniards' physical constitution: "Ello es que, sin largas discusiones ni muchos distingos, habremos de confesar los españoles que físicamente somos de marcada inferioridad a casi todos los demás pueblos civilizados. . . . Nada importaría que los españoles fuésemos de inferioridad física, si ésta no arrastrase consigo cierta flojedad de espíritu" (Without long discussions or many reservations, we, Spaniards, have to confess our marked physical inferiority to all the other civilized nations. . . . It would not be of interest that we, Spaniards, were physically inferior, if this inferiority did not also include a spiritual weakness).[54] Racial and geographical effects on the national character could consequently be synthesized in four negative traits: ignorance, excess of imagination, laziness, and, above all, lack of concern for the motherland.

Ignorance, according to Mallada, was the result of many factors, including the Church's ideological repression, absolutism, and people's lack of initiative, all of which created the need for a constant importation of ideas, customs, and goods from northern Europe. This aspect was particularly problematic: "Miremos en torno nuestro, penetremos en los dorados salones de las familias mejor acomodadas; todos los signos de riqueza, todo lo que es magnificencia, todo lo que denota un trabajo caro y bien recompensado, todo ello es extranjero. . . . ¿No es esto ya una doble señal de nuestra pobreza?" (Let us look around us, let us penetrate affluent families' elegant rooms; there, every single sign of wealth, everything that denotes magnificence, everything showing an expensive and well-compensated job, all of it is foreign. . . . Is this not already a double sign of our poverty?)[55] The opposition between wealth and poverty operates here as a mirror of the local–foreign dichotomy, highlighting the contradictory nature of modernization. In similar fashion, the *excess of imagination* resorts to a traditional vision of Spain, particularly of Castile, anchored in the belief that there was a transhistorical national personality that responded to specific geographical conditions. Using Don Quixote as an archetype of these attributes, Mallada shows that the tendency to idealize reality was the result of the arid landscape—the environment's monotony led people to neglect reality and privilege imagination. Fantasy thus became one of the most negative traits of the national psychology as it obstructed the possibility of accurately assessing the real world: "Para todas las clases sociales existe entre nosotros un defecto que me permitiré expresar con una sola palabra: la fantasía: seducidos por todo lo poético, queremos huir de la prosa de la vida . . . y ¡pobres de nosotros! La prosa de la vida es la realidad" (In all social classes there is among us a defect that I will allow myself to express in one word: fantasy. Seduced by everything that is poetic, we want to escape the prose of life . . . and woe is us! Life's prose is reality). To which he later adds, "¡Sí! La fantasía, la loca fantasía es nuestro principal defecto; la fantasía convierte en un verdadero laberinto la administración pública; la fantasía nos hace ser los mayores proyectistas y los más holgazanes de Europa" (Yes! Fantasy, crazy fantasy is our main defect; fantasy turns public administration into a real labyrinth; fantasy makes us the biggest projectionists and the laziest people of Europe).[56]

Besides ignorance, fantasy, and laziness, Mallada was also concerned with the Spaniards' *lack of practical sense*, which, in his view, directly affected educational, scientific, and technological development: "Sin contar los motivos inherentes al bajo nivel industrial de nuestro país, la enseñanza práctica y las aplicaciones de las ciencias se hallan en espantoso retraso, ya porque a ellas hemos venido demasiado tarde, o porque recibimos la luz reflejada y no directa, o por la fantasía engañadora que nos acompaña hasta el sepulcro" (And that is not taking into consideration the inherent consequences produced by the low level of industrialization in our country; practical instruction and application of sciences are in horrible underdevelopment, either because we started paying attention to them too late, because we only received the reflected light instead of the direct beam, or because of that tendency to rely on the deceiving fantasy, a propensity that will accompany us to the grave).[57] Although the essay includes a denunciation of the country's spiritual and material shortfalls, Mallada does not propose clear solutions to these problems. Instead, and in line with Charles Lyell's foundational principle of geological Uniformitarianism, he accepts the immutability or steadiness of the national character, even when it was considered a harmful trait for the progress of society.[58]

But Mallada's social foundry entails much more than a negative depiction of national identity. In recognizing the political achievements of the 1868 Revolution, for example, his assessment underscores the unique opportunities that this event yielded for revamping society: "Cierto es que las sacudidas que despertaron al país por el robusto brazo de la Libertad, al quitarle el pesado yugo del absolutismo y de la intolerancia religiosa, le guiaron hacia la senda del progreso; . . . sin las reformas liberales de Europa entera, España hubiera seguido con su inquisición y con sus frailes, con sus reyes absolutos y con sus apergaminados señoríos" (It is true that the shaking with which Freedom's strong arm awakened the country, by removing absolutism and religious intolerance, pushed it into the path of progress. . . . Without the liberal reforms that took place throughout Europe, Spain would have continued focusing on its inquisition and its friars, and on its absolute kings and their parchment-like estates).[59] Liberal reforms were also responsible for the emergence and consolidation of a new entrepreneurial class, which Mallada refers to as "aristocracia de negocios" (business aristocracy).[60] This class, in contrast with its predecessor (traditional aristocracy), had a better understanding of the social, economic, and political challenges of industrial development. One of the obstacles to modernization, however, was the persistence of old forms of political administration and social regulation (estates, class immobility, strict religious supervision, and so forth), which, from Mallada's perspective, particularly affected the mechanisms that the country employed to exploit its natural resources: "Es lamentable y doloroso que en estos nuestros tiempos, de tan grandes y rápidos adelantos, casi lo mismo que sucede en los países que jamás acaban de salir de la barbarie, las cuatro quintas partes en valores de los minerales producidos en España se exportan como materias primeras, para ser beneficiados en el extranjero, donde sacan de ellos las principales utilidades" (It is unfortunate and painful to see how in our time, one of great and fast development, four-fifths of mineral production in Spain—similar to countries where barbarism persists—is exported to be used overseas, where foreigners extensively profit from it).[61] This propensity to take for granted national wealth could also be attributed to specific traits of the national character—the tendency to privilege the personal over the collective clashed with the productive principles that defined the industrial paradigm. Whereas industrialized countries around Europe exploited their natural resources to reinvest the profits in their own scientific and technological development, in Spain only a small portion of the possible benefits remained in the country, where, instead of being used to improve the local conditions, they became part of personal fortunes. It was this distortion of the advantages of economic modernization for the benefit of the elites that, in Mallada's view, had produced the irreparable decomposition of the country.

In Mallada's social foundry the symptoms of the ailment afflicting the nation are clearly identified: "Son entre nosotros males irremediables el desbarajuste administrativo, la impotencia y la incapacidad de los Gobiernos, por un lado; la apatía y la ignorancia por otra; la falta de patriotismo, por todas partes y todos los asuntos" (Among us there are irremediable ailments: on the one hand, bureaucratic disorganization, governmental impotence, and administrative incapability; on the other, total apathy and ignorance. A lack of patriotism everywhere and in all matters).[62] One of the suggested ways in which this situation could be alleviated consisted of undertaking a sort of endo-colonization—instead of engaging Spain in occupation campaigns overseas, like the ones carried out by other European countries, the ruling classes needed to take control of the population displacements and relocations that industrial expansion had provoked. An adequate administration of the local work forces and the consolidation of a business class with a solid patriotic sense would indeed prevent the internationalization of local industries and contain the country's economic colonization. To illustrate the current level of disorganization and mismanagement of the country's human capital, in his essay Mallada points to alarming statistics of labor expatriation:

> No os admiréis de que todos los años, del país donde dicen que faltan brazos y sobra inteligencia, emigren más de 25.000 españoles a países que no son nuestras colonias, en tanto que las cuatro quintas partes de nuestras minas de importancia, no pocas fabricas y muchas fuerzas activas que imprimen movimiento a las transacciones mercantiles se hallan en poder de los extranjeros, para quienes venimos a ser unos . . . indígenas.[63]
>
> (Do not be surprised to learn that every year, from a country where it is said that hands are needed and there is an excess of intelligence, more than 25,000 Spaniards emigrate to countries other than our own colonies; while, at the same time, four-fifths of our mines, not fewer factories, and many of the enterprises behind the trade's movement and dynamism are in control of foreigners, to whom we don't represent more than . . . uncivilized natives.)

To mitigate the symptoms of this ailment, Mallada suggests that the government focus on promoting science and education, on completing a detailed study of the country's geography and natural resources, and on doing a serious analysis of the population's dynamics of displacement and relocation.

Since geology and science in general were such important tools for the evaluation and treatment of the country's problems, Mallada saw with profound pessimism and great preoccupation their precarious development in the country. In a speech at the Royal Academy of Sciences in 1897, he lamented: "Pero, si vuelvo la vista atrás y pienso al propio tiempo en su inmediato porvenir, no sé por qué dolencia de mi espíritu o congoja de mi corazón veo el final de este siglo como si fuese la caída de una tarde fría y destemplada de invierno" (But if I look back, and at the same time think about the immediate future, I am not sure what spiritual malady or angst in my heart makes me see the end of this century as if it were the end of a cold and unpleasant winter afternoon).[64] Spain's defeat in the Spanish-American War in 1898 would in fact corroborate Mallada's fears by exposing the country's backwardness not only in terms of military capacity but also with regard to scientific development. Many intellectuals and politicians of the period even went back to revisit the heated *Polémica de la ciencia española* that had taken place decades earlier to see if they could find answers there. The argument, according to which Spain lacked the tradition, talent, and resources to produce science and technology, started receiving attention once again, becoming integral part of the debate on educational reform. Proponents of revamping higher education called attention to an evident reality: there were no minds in Spain able to understand the secrets of nature and use them to improve the country's economy.

Mallada's address at the Royal Academy of Sciences has multiple points of contact with Echegaray's 1866 speech. Both scientists frowned on the conditions of science in Spain; however, while Echegaray focused on mathematics and held religious devotion responsible for the backwardness of the country, Mallada's claims centered on geology and looked back at the nineteenth century's long history of military interventions and political tensions as the main cause for the nation's scientific paralysis. Mallada's measuring of Spain's development in relation to other European countries thus reaffirmed the notion of peninsular exceptionalism that he ardently criticized in *Los males de la patria*: "Europa entera avanzó rápidamente en todos los ramos del saber humano, en miles de invenciones y descubrimientos . . . mientras España seguía estacionada, marcándose su atraso de año en año con mayores diferencias: sobresalían entre nosotros enjambres de políticos y de literatos, y apenas se veía un hombre científico" (Europe advanced fast in all fields of knowledge, in thousands of inventions and discoveries . . . while Spain was stuck, and its backwardness got more pronounced every year with bigger differences: swarms of writers and politicians stood up among us, but only a handful of scientist remained visible). Luckily, he continues, "la geología fué entre todas las ciencias una de las que más pronto salieron de tan afrentoso marasmo y abandono, lo cual fué debido á la industria minera y á los fundadores del Cuerpo de Ingenieros de Minas, quienes, de grado ó por fuerza, tenían que fijar sus miradas en los caracteres petrológicos y estratigráficos de nuestras montañas" (geology was, among all sciences, one of the first ones to end the outrageous atrophy, a change that was possible thanks to the mining industry and the founders of the Corps of Mining Engineers, a group of people that by will or commission decided to study the petrologic and stratigraphic nature of our mountains).[65] Not only did geology constitute an essential tool for the development of mining and, consequently, of industrial modernization, but as a symbolic referent it also functioned as the basis for a social foundry whose breadth could go *beyond the surface* of the national problems.

Science and innovation, moreover, were collective endeavors. Practical inventions, for example, were the result of the combined efforts of a wide and active community of educators, scientists, and technicians who, working from different fields, added the necessary elements to turn ideas into tangible applications. Scientists' and scholars' social repercussions, though less concrete, would always be significant insofar as the knowledge they produced could be disseminated:

> Con sus microscópicos descubrimientos, con sus diminutas observaciones, con sus prolijos detalles, acumulada labor tan inmensa por tantos naturalistas, la ciencia crece y se extiende de un modo prodigioso, á la manera que se forman grandes islas madrepóricas en el Océano por pequeñísimos coralarios que, agrupados en innumerables miríadas de individuos, milímetro á milímetro, generación tras generación, una colonia junto á otra, elevan extensos y sólidos archipiélagos.[66]
>
> (With its microscopic discoveries, with its tiny observations, with its exhaustive details, science grows and extends in prodigious ways once numerous naturalists accumulate such an immense amount of work, in the same way that large madreporarian islands are formed in the ocean out of minuscule stony corals. Grouped in uncountable myriads of individuals, millimeter by millimeter, generation after generation, one colony alongside another, these organisms raise vast and solid archipelagos.)

For Mallada, collaborative work was essential to consolidating a national scientific culture composed of strong institutions, adequate curricular plans, and wide state support. By offering a complete scientific revision of the national wealth, geology was the only discipline in Spain that was close to achieving these goals.

Mallada's social foundry resorts to a particular repertoire of images in order to describe and examine national identity. Both in *Los males de la patria* and his speech at the Royal Academy of Sciences, a persuasive literary rhetoric is deployed to generate an accurate diagnosis of the country's ailments and a compelling prescription for the possible remedies to treat them. In the previous quoted passage, for example, Mallada uses a double analogy that incorporates a physical concept—energy accumulation and transformation (in this case the sum of scientific efforts to invigorate social development)—and a biological reality—the consolidation of social structures assimilated into an oceanic phenomenon. This allegorical exercise relied on his audience being a scientific community able to visualize the complex images and understand the specialized language drawn from geology, physics, and microbiology. Thus Madreporaria, an order of microorganisms responsible for coral formation, represents the imperceptible but persistent contribution of scientists to national modernization. Mallada's use of oceanic images is far from being arbitrary; the word *Madreporaria*, for instance, is formed from a combination in French of the words *madré* (skillful, clever) and *pore* (porous, pervious), which underscores the extraordinary capacity of minuscule organisms to produce large and complex structures, in the same way that little scientific efforts formed the scaffold of modernization. Scientists, like "stony corals," had to group and work together to create vast networks of knowledge that supported industrial development.

Mallada's rational and systematic study of society entailed a conceptual and a practical dimension. On the one hand, his social foundry was based on a meticulous scrutiny of geological and geographical conditions—its representation of society therefore depended entirely on theoretical models in which geology worked as a lens for evaluating reality. On the other hand, while completing his *Proyecto de una nueva división territorial de España* (Project of a new territorial division of Spain) (1881), Mallada received a direct experience of urban and rural dynamics. After assessing the country's administrative organization, he coincided with Estasen's interpretation that Spain's backwardness was in part the result of an uneven distribution of resources: "Y porque España es más pobre de lo que se piensa y porque España está más atrasada que el resto de Europa; en Madrid, donde se consume y prospera lo que en las provincias se halla de sobra; en Madrid, de donde irradia la ilustración que en las provincias se recibe, en Madrid está la causa de todos lo males, según los frenéticos partidarios del provincialismo" (Because Spain is poorer than what is thought, and because it is more backward than the rest of Europe, in Madrid, where everything that prospers in the provinces is consumed; in Madrid, the source of the enlightenment that the provinces received; in Madrid, according to the advocators of provincialism, lies the cause of all the national problems).[67] In his field work, Mallada saw how the country operated almost as a colonial structure whose centralism implied an obstacle for social and economic progress. More than an actual shortage of material wealth (represented in natural and human resources), poverty was the cause and consequence of this tension between center and periphery that had altered the distribution of intellectual and cultural capital and generated a relationship of internal dependency.

While the social foundries explored here suggest that the national problems can be linked to deficiencies in education, their prognosis focused on evaluating the economy and proposing new perspectives on the country's geological and geographical configuration. Indeed, Echegaray, Estasen, and Mallada propose original approaches to make an effective use of natural and human resources, projecting a future in which all Spaniards, regardless of their ideological or political views, would work together on creating the conditions for a particular kind of modernization—one in which national identity would not be in conflict with technological progress. However, the problem of education deserves particular attention as it offers yet another angle on industry's cognitive transformation. By conceiving models of scientific instruction, institutional organization, and curricular structure using industrial images, principles of operation, and processes as referents, it was possible to incorporate education into the country's projections. Chapter 3 explores essays that deal with the challenges posed by this integration to show how the material needs of fin-de-siglo Spain led to a debate on the practicality of knowledge that somehow mirrored this period's tensions between tradition and progress.

CHAPTER 3

The Educational Engine

As different politicians, scientists, and educators attempted to understand the country's inability to consolidate its scientific culture at the end of the nineteenth century, various educational models were proposed to align learning goals and plans of study with the country's industrial needs. Notions of cultural exceptionalism, used by many intellectuals of the time to justify the country's backwardness as just another trait of the Spanish character, seemed to confirm the difficult reconciliation in the political and cultural arenas of the two ideological perspectives that dominated the national debate: traditionalism, with its strong attachment to Catholicism and social stratification, and progressivism, with its links to positivism and liberal thinking. Under the pressure of confronting their identity against objective reality, *scientific mentors*' proposals to reform higher education became complex spaces of diagnosis, prescription, and prognosis of the national situation. This chapter explores two different areas related to these problems. On the one hand, it deals with scientific concepts, their reception, and their assimilation in *fin-de-siglo* Spain. On the other hand, it analyzes how the complexity of these discoveries, as well as their dissemination, provoked questions about Spaniards' capacity of innovation and promoted a (re)evaluation of the system of higher education.

In contrast to more developed European countries, in Spain those seeking the advancement of original scientific ideas or the invention of new technologies encountered multiple obstacles. This situation was in part the result of an educational system that privileged practical over theoretical knowledge. Access to the highest levels of instruction, on the other hand, was also limited to a small portion of the population—only a privileged elite was able to pursue the necessary degrees to be in charge of the country's modernization.[1] Furthermore, lack of technical resources and deficiencies in the structure of the programs of study dedicated to the sciences made it necessary to import knowledge and expertise from other countries, thus preventing the consolidation of a local scientific tradition. Concerned with the historical causes of this problem, for example, in his 1866 speech at the Royal Academy of Sciences Echegaray noted: "No es esta . . . la historia de la ciencia en España, porque mal puede tener historia científica pueblo que no ha tenido ciencia [sino sólo] látigo, hierro, sangre, rezos, braseros y humo" (This [his speech] is not the history of science in Spain, for as a country that instead of being exposed to science has been subjected to whips, iron, blood, praying, stakes and smoke cannot have scientific history).[2] In fact, dependency on traditional institutions, such as the Church (hence the reference to inquisitional practices) or the aristocracy, had been one of the main obstacles to the country's progress. The scientists and educators studied here not only acknowledged this reality but also worked on counteracting its consequences by promoting curricular reforms and organizational changes. The creation of institutions for the advancement of science, like the Junta para Ampliación de Estudios e Investigaciones Científicas (Council for the Expansion of Scientific Research and Studies) in 1907, is an excellent example of these initiatives.[3]

The widespread presence of foreign ideas, in any case, was useful not only for revealing profound scientific and technological deficiencies in the country but also for introducing authors to new perspectives on education that had proved to be successful elsewhere. If, as journalist, educator, and reformist intellectual Ricardo Macías Picavea contended, Spanish culture was only a "second-hand culture," created almost entirely to emulate other countries' achievements,[4] why then should it not take advantage of some of these "second-hand" successful experiences to foster national education? Thermodynamics, as noted in chapter 1, was one of those attainments—a branch of *physics* that had created a revolution on its own by offering conceptual tools for cultural and economic analysis. Following this logic, other scientific fields turned as well into useful referents for the evaluation of reality: *chemistry* and *mathematics*, for example, provided valuable elements to approach social problems. By reviewing some aspects concerning the instruction, study, and learning of these three disciplines, in what follows I explore the concrete proposals to reform the national educational system of scientists and educators Gumersindo Vicuña, Laureano Calderón, Zoel García de Galdeano, and José Rodríguez Mourelo. These authors adapted their *social foundries* to the conceptual and practical aspects of their specialty and came to valuable, original conclusions. Their main preoccupation was the development and implementation of organizational structures and systems of instruction that helped the country generate a local scientific culture while, at the same time, producing practical and palpable results that actively contributed to its belated modernization.

Gumersindo Vicuña and the Engineering of Higher Education

In 1875, amid the profound agitation that followed the promulgation of Orovio's infamous decree that deprived scholars of academic freedom, engineer and educator Gumersindo Vicuña threw himself into the public sphere with an unconventional proposal. He considered that one of the main problems with the educational system in place was the students' automatic admission into science programs without previous examinations to determine their potential. Vicuña wanted to advance a strategy to change the mechanisms of selection used by universities across the country. In the opening remarks of his lecture at the inaugural session of the 1875–76 academic year at the Universidad Central he explained, "El defecto capital que presentan nuestras Facultades de ciencias, aquel contra quien claman unánimemente todos sus profesores, el que es capaz por sí solo de esterilizar los esfuerzos de estos, por grandes que sean, consiste en que puedan comenzar los alumnos a estudiar en la Facultad con sólo haber probado [*sic*] la segunda enseñanza" (The main problem with our science departments, against which all professors unanimously complain, the one that is by itself able to make unproductive their efforts no matter how big these are, is that students are able to start studying in college after only having completed high school).[5] The talk's title, "Cultivo actual de las ciencias físico-matemáticas en España" (Current promotion of physical and mathematical sciences in Spain), narrows Vicuña's criticism to curricular problems in the programs of physics and mathematics, two disciplines particularly relegated to their practical uses in engineering but little studied as theoretical sciences. The increasing demand for qualified workers, supervisors, and engineers to work in industry gave justification for the institutional emphasis on technical rather than theoretical instruction. Curricular structures based on this recommendation, however, had prevented the cultivation of actual scientists and, in general terms, the necessary reorganization of the system of higher education.

Vicuña had completed degrees in engineering and sciences before becoming a professor at the Universidad Central de Madrid, where he taught physics and mathematics. As part of his academic position, he traveled to various European countries, where he saw how different educational systems contributed to the progress of science and the development of society. Some of the impressions collected in these travels, as well as his own experience as a scientist and educator, proved to be crucial during his political career as a legislator representing Biscay, a post he used to advance multiple projects to improve the conditions of science and industry in the region. After studying the structure of the educational apparatus in France, Belgium, and Germany, he advocated for the creation of three distinctive sections at the high school level—the purpose was mapping students' talents and interests before they chose a particular major or field of concentration in college:

> Por esto, y partiendo siempre del estado de nuestro país, sería preferible refundir las Escuelas especiales en tres grupos; el primero para profesar las *artes de construcción*, de donde salieran los ingenieros de caminos y los arquitectos; el segundo para las *artes industriales*, donde se formáran [*sic*] los ingenieros mecánicos, quimicos y mineros, y el tercero para las *artes gráficas*, que equivaliera á las escuelas actuales de agricultura y de montes. Éstas son las tres divisiones que aconseja el estado actual de las ciencias de aplicación, prescindiendo de consideraciones de órden [*sic*] distinto, algunas en pró de este sistema y otras en contra, que son las que hoy prevalecen.[6]
>
> (Because of this, and of course taking into account the current state of our country, it would be preferable to reorganize special schools into three groups: the first devoted to the *construction techniques*, where architects and road engineers would complete their trainings; the second dedicated to the *industrial techniques*, where mechanical, chemical, and mining engineers will be instructed; and the third one for the *graphic techniques*, which would be equivalent to the current schools of agriculture and forestry. Without taking other approaches into consideration—some of which are in favor and others against this system, with the latter ones prevailing today—these are the three divisions that would best fit the current state of the applied sciences.)

This subdivision clearly aimed at promoting disciplines that could have immediate effects on the country's economy. In fact, Vicuña envisioned a university where resources were allocated into multiple engineering schools covering areas of development such as construction, industry, or design. While the development of natural sciences was not imperative within this project, the creation of an adequate educational structure to facilitate the country's modernizing efforts was deemed essential for the future. Rather than being formally taught, science could then be communicated through the popularization genre, a cultural field to which Vicuña copiously contributed through the publication of texts such as *Elementos de física al alcance de todo el mundo* (Elements of physics accessible to everyone) (1874) and *Manual de física popular* (Handbook of popular physics) (1878).

Vicuña's emphasis on the practical aspects of science, however, did not imply that scientific research and theoretical analysis were elements excluded from his social foundry. On the contrary, since science was the only mechanism through which humans could gain control of nature, in his view modernization would be impossible without it: "Uno de los caracteres distintivos de la moderna civilización comparada con la antigua, es precisamente el empleo de los conocimientos científicos para domeñar la naturaleza" (One of the distinctive traits of modern civilization, if one compares it with the old one, is precisely the use of scientific knowledge to tame nature).[7] If one of the main goals of restructuring the system of higher education was to have a better distribution of talent so that students could eagerly contribute to the country's projects of modernization, scientific research needed to be part of the curriculum. Vicuña took advantage of his own experience as a visiting scholar at different European universities to evaluate the ways in which other educational systems had balanced their practical and theoretical needs. From his perspective, systems like the ones in place in Belgium or Spain, in which there was no distinction between technical institutes and universities (the natural place for research and theoretical work), hindered the promotion of science, as state resources for education could not be properly allocated. The French system, on the other hand, was too technically oriented; a system that might work well in a country with a significant amount of scientific research and a strong tradition of scientists able to develop new and innovative ideas. Finally, the German structure offered an acceptable balance between theory and practice and might serve as a model for future educational reforms in Spain.

Reflecting on these three possibilities, in his talk at the Universidad Central Vicuña criticized the lack of state investment and the scarcity of monetary resources dedicated to public education in Spain: "Desde luego es justo hacer notar que la causa principal de nuestro relativo atraso en estas cosas [la educación pública], arranca de la pobreza y apuros de la riqueza pública y del corto número de grandes fortunas individuales" (It is indeed fair to note that the main cause of our relative backwardness in these matters [public education] derives from our public finances' shortages and predicaments, and the reduced amount of big individual fortunes).[8] Immediate consequences of these limitations were the absence of libraries, research institutions, and scientific associations. A revamping of the educational system therefore required working on two distinct fronts: strengthening instruction and generating conditions for research. In order to achieve these goals, technical preparation and formal learning needed to be separated by means of creating specialized schools or institutions that would assume the role of promoting practical or applied knowledge. Universities, for their part, could continue serving as repositories of theoretical erudition. Vicuña's proposal was based on a transformation of the current system to accommodate this subdivision; he acknowledged that the cost of initiating a whole new system was excessive even for countries with better economic circumstances than the ones in Spain. Considering this, he added, "Las reformas útiles en la enseñanza son las que se apoyan en lo existente y lo mejoran; no las que destruyen por completo un régimen y hacen tabla rasa de lo antiguo para implantar lo nuevo" (Useful reforms in education are the ones that support what is already in place and improve it; not the ones that completely destroy a system, wiping the slate clean to impose something new).[9] Vicuña's social foundry therefore suggests a particular engineering of the educational structure that, without disregarding tradition, profited from Spaniards' natural disposition to progress.

Expanding on Vicuña's yearning for an educational system in which practical and theoretical knowledge could be balanced to strengthen industrial modernization, a decade later Laureano Calderón addressed the problem of curricular reform to stress the need for bonding spirituality and materialism in the project of consolidating a *national* scientific culture.

Laureano Calderón and the Educational Power of Industrial Imagination

By 1884, the year in which Laureano Calderón gave his opening lecture at the Sección de Ciencias Exactas, Físicas y Naturales (Division for the Physical, Natural and Exact Sciences) of the Ateneo of Madrid, basic notions of theories such as thermodynamics or evolution were widely known and had made their way into the collective imagination. As already noted, in this period scientific terminology and theories circulated as common currency to describe and even justify social phenomena. However, this unprecedented convergence of science and popular knowledge was not exempt from producing misinterpretations, which, in turn, had the effect of exacerbating conservatives' fears and qualms regarding scientific and technological modernization. The ironic tone in Calderón's introductory remarks at the Ateneo actually points to this (mis) appropriation and informality—science had turned into just another trending topic of discussion at social gatherings, café conversations, or political meetings:

> Ya no hay jurisconsulto en agraz, ni teólogo en barbecho, ni poeta trascendente, ni dama elegante, ni aspirante á diputado que, evocando los recuerdos del último artículo de un periódico de los llamados político-literarios, no formule, entre las amenidades de la tertulia de confianza, estupendas conclusiones sobre la descendencia del hombre, los períodos geológicos ó la pluralidad de mundos habitados.[10]

> (Today there is no aspiring law clerk, unprepared theologist, important poet, elegant lady, or parliament contender who, evoking their reading of the last article included in one of those variety newspapers, is not able to present, surrounded by the amenities of their habitual social parties, unexpected conclusions about the descent of Man, the different geological periods, or the plurality of inhabited worlds.)

This access to scientific ideas, however, did not imply that Spanish society was properly trained to potentialize modernization and actively contribute to social change. A thorough understanding of science, which was possible only through a structured education, was essential to creating stable conditions for technological innovation. Calderón saw society as a malleable structure that could be molded using systematic procedures, a notion he carefully articulates in his address at the Ateneo and complements in other talks. In these interventions, he dissects the system of higher education to present a formula that in principle would help generate the kind of professionals an industrialized nation needed.

Calderón's social foundry, consequently, focuses on labor, education, and scientific development as some of the ingredients with which to forge the foundations of national modernization. For him progress was the result of both natural and divine forces—a mixture of scientific materialism and religious spirituality. In order to deal with this duality, he resorted to Krausism, a doctrine of which he was a prominent proponent. Indeed, Calderón had been a founding member of the Institución Libre de Enseñanza and a direct disciple of Julián Sanz del Río. From this standpoint, Calderón pioneered the teaching of biochemistry in the Universidad de Santiago, where he held a professorship. After Orovio's 1875 reforms, Calderón was removed from his position at the university, a situation that forced his relocation to Germany, where he had the chance to continue his studies in the field of crystallography.[11] It would be overseas that he consolidated his career as a researcher and saw firsthand how the German system of higher education worked, an experience that would allow him to assess and reflect on the most serious deficiencies of the Spanish educational apparatus.

Whereas in Germany, as Calderón notes in his reflections, curricula were perfectly structured to balance pure and applied sciences, in Spain state administrators privileged practical and applicable uses of science over theoretical knowledge, thus relegating research and teaching to a secondary role. Addressing this issue, in his lecture at the Ateneo in 1884 he decried the consequences this practice had on the social perception of science and scientists: "No es del todo inexplicable esa especie de anatema que pesa sobre los naturalistas. A fuerza de mostrar al mundo que son capaces de vivir sepultados entre piedras, plantas, animales, retortas y crisoles, sin que semejante voluntario ostracismo los lleve a fin alguno práctico, ni siquiera á ser ministros, han dado muchas gentes en suponerlos poco merecedores de la estima social" (Common distrust of naturalists is not completely unfounded. By means of showing the world that they can live surrounded by stones, plants, animals, retorts and crucibles, with such voluntary ostracism not serving any practical end, not even the one of becoming ministers, many people have assumed that they don't deserve social esteem).[12] Calderón considered that science had an important role in society. Consequently, scientists needed to become active agents in the projects of national modernization, an idea he would bring once again into discussion in his talk at the opening of the 1892–93 academic year at the Universidad Central in Madrid:

> Celebremos, pues, hoy la fiesta universitaria: proclamemos en buena hora el derecho con el cual aspiramos á representar el organismo docente como

> encarnación viva de un fin del Estado; pero, al propio tiempo, dirigiendo nuestras miradas más allá de este recinto, reconociéndonos todos miembros de una sociedad, parte á su vez de la familia humana, solicitemos del medio social el concurso inexcusable sin el que la vida universitaria degeneraría en ascetismo intelectual abstracto é idealista.[13]

> (Let us celebrate today the university: let us proclaim, on this appropriate occasion, the right we want to have of representing the whole educational apparatus as an active embodiment of the state's mission; but, at the same time, and looking beyond this room, seeing ourselves [scientists] as part of society, which in turn is a member of the human family; let us demand from that society the inexcusable consensus without which academic life would degenerate into an abstract and idealistic intellectual asceticism.)

Depicted as living forces, educators' contribution to national goals was contingent on the social nature of their roles as scientists and researchers. The main function of universities and other educational institutions in that context was to provide an appropriate space to perform these activities. Industrial imagery is subtly integrated into the speech through the notions of "unity" and "degeneration." From this perspective, it was the productive relationship among members of the academic community that would prevent its disorganization and, consequently, its exhaustion—both, situations that would arise from privileging practical over theoretical knowledge.

Beyond discussions about its practicality or impracticality, however, science could be an innovative and effective tool for understanding society. In fact, when discussing what he saw as the main social challenges of the country, Calderón explicitly reflects on this possibility: "¿No parece indudable también que los movimientos sociales deben producirse según leyes fijas é inmutables? Pues algún día podrá constituirse una *mecánica social*, una política tan segura en sus principios como la mecánica de los astros, con sólo aplicar á la vida de los pueblos las doctrinas y principios de la dinámica" (Is it not evident that social activity should obey fixed and immutable laws? In that way, someday, by just applying the doctrine and principles of dynamics to people's lives, it will be possible to develop a *social mechanics*, a set of rules so accurate in its principles as the ones applied to the mechanics of the stars).[14] As conceived here, the goals of Calderón's "social mechanics" included a complete understanding of the tensions and displacements governing modernization and the ability to intervene in and regulate the transformation of society. His main concern was the country's backwardness—persistent weaknesses in education, continued lack of scientific developments, underlying technical dependency, and ongoing social unrest were just symptoms of a bigger social ailment. Exalting science and education as answers or remedies to these problems thus served multiple purposes. First, it called attention to the need to focus on the future rather than the past. Second, it exposed the precarious situation of the country in terms of scientific development as one of the causes for its delayed modernization. Lastly, it encouraged political administrators to incorporate scientific instruction into their projects of educational reform. In Calderón's view, science was not something to be afraid of, but something to draw on; that was why it was imperative to evaluate its existing condition on the peninsula.[15]

Yet, as Calderón reported, the "current state" of science in Spain was not satisfactory. Another obstacle to consolidating the country's scientific culture was people's resistance to change. Progress was, in general, seen as an evil, a negative force that could disarticulate society by changing its trajectory. Modernization's social transformations, for example, were perceived to have eroded the moral values on which identity had been built; rejecting these changes was therefore expected. Since these adjustments seemed to threaten their privileged place in society, the Church, the aristocracy, and certain sectors of the bourgeoisie were also hesitant to support what they, in many cases, considered necessary changes. In the name of tradition, many attempts to reform the political and educational systems or to alleviate the social tensions arising from labor exploitation were rejected, delayed, or ignored. At the Ateneo, Calderón responded to this situation, remarking, "Caballero de su destino, se lanza el hombre á la lucha de la Vida, y a cada paso vacilante que da, solicítanle en opuestos sentidos la tradición y la crítica racional" (Owner of his own destiny, man throws himself into the fight of living, and with every hesitant step he takes, tradition and rationality pull him in opposite directions).[16] As a Krausist thinker, Calderón acknowledged that humans were in control of their own destinies and recognized that, in a country like Spain, attempts at advancing could always be thwarted by the force of tradition. The image depicted here inevitably evokes a force diagram (one of the most useful tools in physics), exemplifying once again Calderón's interpretation of society as a mechanical system.

Calderón's social foundry, nonetheless, transcends the problem of *forces* and the *unity of forces* (common topics in many scientific studies of this period) to rely instead on the two laws of thermodynamics—principles that were thought to encompass all physical phenomena.[17] When comparing the roles different scientists played in the forging of knowledge, for example, he deems all scientific work to be dependent on those laws: "—Yo, dirá el físico, he seguido las huellas que hombres ilustres me trazaron y he resumido toda la ciencia en dos principios, tan admirables por su sencillez como fecundos en sus maravillosas consecuencias" (I—the physicist would say—have followed the steps traced by illustrious men and have encapsulated scientific inquiry in two principles, as astonishing in their simplicity as they are productive because of their extraordinary consequences).[18] Although these two principles, as noted, became essential to understanding social struggle in theoretical terms, Calderón's social foundry also includes an empirical component. Assuming that social phenomena could not be explained entirely with abstract theories, he thought scientists needed to incorporate their practical experience into their interpretations. The key was to focus on society's dynamic character, as he remarked in his speech at the Universidad Central: "Las constantes de nuestra experiencia solo en determinadas circunstancias pueden ser consideradas como tales constantes; y para fijar su valor peculiar en cada caso se hace indispensable obtener datos que nos consientan conocer cuál sea aquel en las condiciones de la experiencia misma" (Only in specific circumstances the fixed principles of our experience can be considered as such; and to set their particular value in each case, it is indispensable to obtain data that allow us to measure it within the conditions of experience itself).[19] If society were a laboratory, then the forces, energy, and dynamism of its transformations could be measured through experimentation; knowledge of scientific methodologies (data collection and analysis) and proficiency in its language and underlying principles were thus necessary skills for dealing with industrialization's social change.

The rhetoric underlying Calderón's social foundry derives from processes of generalization embedded in the epistemological structure of the scientific method—observing, questioning, predicting, hypothesizing, collecting data, testing, and theorizing thus appeal to the idea of curiosity as one of the main engines of science. In his speech at the Universidad Central, for example, Calderón shows how rationalization and analysis had replaced speculation: "La razón humana ha presentido siempre que deben existir principios generales que coordinen la variedad infinita de las cosas y aspira á conocerlos irresistiblemente solicitada por su propia naturaleza; mas lo que en otros tiempos fué genial presentimiento ó producto de una fantasía poética, aparece hoy como el fruto de la asociación sistemática de leyes confirmadas" (Human reason has always anticipated the existence of general principles that organize an infinite variety of things, and it aspires to know them, always irresistibly inspired by their own nature; but that which in other times represented brilliant intuition or was the product of poetic fantasy, today presents itself as the result of the systematic association that only confirmed laws provide).[20] In the social realm, these "general principles" were used to define, explain, and address phenomena such as the disorganized growth of the working class—its displacements and its relocations—questions that constituted a major concern for both city and state administrators.

From a conservative perspective, these dislocations were unsettling for they seemed to force a refashioning of the traditional social structure. Liberals, in turn, were more preoccupied with the negative effects that labor exploitation had on the workers' morale and with the growing social unrest. Nonetheless, both groups had accepted capitalism as a norm, which implied, as Labanyi explains, that modernization was understood as a necessary force even when it was evident that mobility and other phenomena related to capitalist development could be extremely problematic.[21] Both political sides were thus interested in increasing productivity and accelerating modernization, and the physics of energy offered a justification for the social imbalance inherent in this process. In his speech at the Ateneo, Calderón addresses this problem by characterizing inequality as an indispensable principle for the functioning of society:

> La sociedad, al reconocer la suma de energía de que cada hombre dispone, la proporción en que sus aptitudes se hallan combinadas, el medio en que realiza su existencia, el influjo de la herencia y hasta de las condiciones materiales de su vida; la sociedad, decimos, podrá fijar los límites en que la acción del sujeto haya de desenvolverse, y no exigirá de él responsabilidades que tal determinado individuo no podrá acaso aceptar jamás.[22]

> (Society, by recognizing the amount of energy that each man possesses, the proportions and measurements of the mixture that makes his talents, the influence that heredity and even the material conditions of his existence has; society, we were saying, will be able to set the boundaries within which every subject's action takes place, and won't ever demand from him responsibilities that perhaps he would not ever be able to accept.)

Within a capitalist economy predisposed to the accumulation of wealth and the control of domestic labor, only a social differential would promote the circulation of work. Analogous to the workings of a steam engine or a steel furnace, society required an unbalanced distribution of capacities so that energy could flow; if this differential were to be suppressed, the machine would cease to function. Individuals, however, could use labor as an exchangeable asset to actively participate in the country's economy. Calderón's focus on productivity, and not on the workers, is not surprising within this model: it was founded on the liberal notion that social inclusion—people's involvement in the economy—was positive, even beneficial, as long as it provided the space for trading in labor capacity.[23] Inequality was therefore the unintended but necessary consequence of these transactions that guaranteed the transformation of energy and work into material goods.

Within this model, society was the site where labor was continuously negotiated. Imbalance and dissimilarity were thus not only crucial to creating and keeping up with the dynamism of modernization but also necessary in the measurement and evaluation of progress. Calderón returns to this idea in his lecture at the Universidad Central, emphasizing once again the empirical nature intrinsic to his assessment of reality: "Nuestros sentidos no aprecian ni perciben otra cosa que caídas, diferencias, desequilibrios, desniveles de potenciales; nunca las energías mismas, en medio de las cuales vivimos y nos movemos" (Our senses cannot appreciate or perceive anything but power declinations, differences, unbalances, slopes; never would they be able to sense the very energies wherein we live or move).[24] As mentioned earlier, industrialization's profound impact on perception particularly affected notions of time and space. This cognitive transformation also changed the formulation of ideas and projects related to national cohesion and territorial unification. It was in part this violent dislocation of the senses that allowed scientists to effectively combine theory and practice in their evaluation of social change—as it offered a unique and essential understanding of the country's most urgent challenges, the materialistic nature of this task could be overlooked.

In the talks discussed here, Calderón's reading of society presupposes a rupture between past and present in order to facilitate modernization; this point is particularly stressed in his denunciation of science as a segregated activity in Spanish society. The literary depiction of the encounter between a geologist and a priest in a remote point of the national geography, which appears in the introduction of his speech at the Ateneo, is a good example of this prospect. This characterization concludes with a piece of advice to his readers: "A los que pretendan resucitar ideas rancias, viejas preocupaciones, dogmatismos absurdos y tradiciones añejas, sólo cabe aconsejarles la paciencia y la resignación para aceptar con calma un mundo que se impone" (For those who pretend to revive old-fashioned ideas, old preoccupations, absurd dogmatisms, and ancient traditions, I can only suggest patience and fortitude in calmly accepting the world that is imposing).[25] For Calderón, the unstoppable advance of science and technology was a reality, an imposition of the times. The real question was not if, but when Spain was going to embrace this transformation by assuming its own capacities and possibilities. Only through the exploitation of its potential energy and labor force would the country be able to revitalize its process of modernization. Science and education were the engines of this change; universities and other educational institutions, the foundries of progress.

Knowledge of physics, chemistry, and particularly thermodynamics provided Calderón with a new logic to diagnose the country's educational problems. Furthermore, the widespread use of a physics-based rhetoric that brought energy and work together demonstrates the effectiveness of the cognitive assimilation of industry for understanding the complexities of progress. This dialogue also had reverberations in religious and political matters. It is not surprising, then, that the aristocracy and the Church participated in state politics, pushing legislative changes that affected science and science education to serve their own interests, or that liberals and progressive thinkers espoused contradictory positions, simultaneously denouncing worker exploitation and favoring capitalist growth. Similarly, in this period companies maintained close ties with, or participated directly in, the decisions made by the government with regard to industrial development. In each case, industrialization affected not only how society was understood and interpreted in the present, but also how it was projected into the future.

Calderón's social foundry also makes the case for the inclusion of science in the regular operation of modern society. Transcending the materiality of progress, in that depiction science was one of the main sources of energy to fuel the construction of national identity. That is precisely the conclusion he comes to in his talk at the Universidad Central: "Los pueblos en los cuales la Ciencia como función no posee una vitalidad y una energía propia, carecen de aquellos resortes movidos por una idea nacional que, vigorosamente acentuada, desborda á todas las manifestaciones de la vida" (Countries where science functionality does not have its own vitality and energy lack the necessary mechanisms to animate a national idea, one that, when promoted vigorously, surpasses any manifestation of life).[26] The idea of a national science is projected here as a pivotal attribute of the modern nation. Like other Krausist thinkers and liberal educators of this period, Calderón believed in the value of scientific instruction and the importance of knowledge popularization as central components of modernization. Education was therefore the only way to generate the conditions for developing such a scientific culture. That was the message of both his speech at the Ateneo and his lecture at the Universidad Central. In both cases, the topics of the talks (the state and development of chemistry in Spain), their form (academic lecture or keynote address), and their language (scientific and industrial images, concepts, methods, and vocabulary) converge to present a clear and novel evaluation of society.

At the beginning of the twentieth century, however, Spain had not yet undertaken the necessary structural remodeling of its educational apparatus. The consolidation of a national science in that scenario seemed to be more a dream for the future than a concrete possibility in the present. Yet the urgency to produce palpable, immediate changes in the educational structure continued being part of many political, intellectual, and economic projects. Such was the case of Zoel García de Galdeano's journal *El Progreso Matemático* (The mathematical progress), a platform this mathematician used not only to disseminate the latest discoveries in mathematics but also to discuss and propose a different vision of the system of higher education in Spain. In the next section I discuss his suggestions, highlighting the way in which industry and technology informed his assessment of the country.

Zoel García de Galdeano and the Science of Science Education

In 1891 Zoel García de Galdeano, a mathematics professor at the Universidad de Zaragoza, embarked on a quixotic enterprise. Seeking to make the most important advances in mathematics available in Spain, he founded the first specialized journal of this field in the country and served simultaneously as its editor, contributor, and manager. *El Progreso Matemático*, as the journal was called, also provided a space for criticizing the backwardness of education and the absence of local scientific developments on the peninsula. Three comprehensive essays authored by García de Galdeano and published in his journal from 1893 to 1894, and from 1899 to 1900, are emblematic of this editorial stance. The pieces, entitled

"Estudios sobre la enseñanza y el organismo de la ciencia matemática" (Studies on the teaching and organization of mathematics), "La enseñanza de la ciencia matemática en la universidad" (Mathematics instruction at the university level), and "La matemática y su enseñanza" (Teaching mathematics), invoke mathematics instruction as a paradigm for developing and implementing an entirely new structure of the educational system in the country. In this view, theory and practice—treated as unconnected instances of the educational body—should be correlated yet independent components of higher education, similar to the way in which pure and applied knowledge constitute the core of science. As in Vicuña's and Calderón's social foundries, for García de Galdeano it was higher education's privileging of practical applications over theoretical knowledge that had dramatically limited the possibilities of autonomous scientific and technical development in Spain. He believed in the broad concept of the university, a space where education could provide a compendium of universal knowledge and foster an understanding of both the sensorial and the conceptual worlds. In this context, the teaching of mathematics was instrumental in supplying, on the one hand, the tools to create an abstract representation of natural phenomena, and, on the other, the concepts to understand the highest notions of philosophy.

To address Spain's backwardness and economic disadvantage in relation to other European nations, the essays discussed here insist on the idea that science education should comprise both theoretical and practical matters, including a critical understanding of its philosophical foundations and not only a notion of its utility in industry. This was essential to developing "la ciencia propia" (our own science),[27] as it would provide students with the long needed aptitude for investigation. Through the advancement of these ideas on education, *El Progreso Matemático* managed to become a privileged arena for social criticism, political discussion, and cultural exchange, contributing in that way to the diagnosis and treatment of the country's problems. In this light, García de Galdeano's social foundry can be framed within the ideological limits of regenerationism, where his proposal joined forces with those of other scientists and intellectuals of his time preoccupied with the country's moral afflictions.[28]

The emphasis on the need to develop an adequate educational apparatus called attention once again to the impossibility of forging "our own science" without the ability to foster the distinctive creativity and ingenuity of the Spanish intellect.[29] Affiliated with Krausism, García de Galdeano was of the opinion that the circulation of knowledge was integral to science education. In fact, Krausists endorsed the freedom of the press (one of the achievements of the revolutionary period) to actively participate in what historian José Sala Catalá defines as the first national movement of scientific popularization.[30] As *El Progreso*'s list of subscribers attests, this posture captured the attention of an important number of educators and politicians who strongly believed in the contribution of the journal to establish a national scientific tradition.[31]

The development of scientific cultures around Europe was in part the result of strong and well-established legacies of intellectual inquiry. As Spain did not count with a homologous endowment, new ideas in fields such physics, chemistry, or mathematics had to be imported.[32] While on occasion this impediment made Spanish authors' engagement with novel intricate concepts and theories somewhat problematic, many of these notions certainly altered their understanding of reality.[33] The cultural significance of these epistemological transformations, as cultural critic Andrea Henderson contends, was that they questioned the validity of established models of representation.[34] Mathematics, in particular, impacted the ways in which some aspects of education were reconceived. García de Galdeano's social foundry, for example, incorporates advanced mathematical concepts into the evaluation of teaching and curricular design. Thus, Georg Cantor's theory of sets, a revolutionary advancement in logic, is used as a referent in "La matemática y su enseñanza" to illustrate the correlation that exists between theory and practice:

> La inteligencia se apodera de un sistema abstracto, correspondiente a un sistema concreto; y nos hallamos en un primer caso de correspondencia, que juntamente con muchos otros que obtendremos, permite definir la matemática, en su concepto abstracto, como ciencia de relaciones formales . . . donde lo real, lo ideal, lo imaginario, lo intuitivo ó lo puramente racional existen, con igual validez, como objeto de nuestras investigaciones.[35]
>
> (Intelligence takes over an abstract system that corresponds with a concrete one; and now we have the first case of correspondence, which, in conjunction with many others we can find, will allow us to define mathematics, in its abstract meaning, as a science of formal relations . . . where the real, the ideal, the imaginary, the intuitive or the purely relational exist, with the same validity, as subjects in our research.)

The proposed pedagogy here is based on the progressive construction of correspondences between reality and representation. As in Cantor's model, numbers can be assimilated as the irreducible elements of any combination in nature. Concepts of equality and subordination, among others, can then be expressed as abstractions of the logical relations that exist among these numbers. For, as García de Galdeano emphasizes, "la matemática tiene parcialmente su realización en la Naturaleza; en ésta simultáneamente se realizan las leyes del número aplicadas á variedad de objetos; espacio, tiempo, fuerza" (mathematics is partially realized in Nature; in it, the laws of numbers and their application to different objects [space, time, forces] take place simultaneously),[36] learning basic notions of arithmetic provided the conceptual structure needed for understanding more advanced mathematical principles, like those embedded in the laws of thermodynamics or other physical notions that could later be applied in industry.

Besides dealing with mathematics instruction, and on occasion responding to the country's political and social unrest, the analyses published in *El Progreso* were also persistent denunciations of the local scientific unsophistication. Through translations, bibliographical reviews, and biographical sketches, the journal introduced in Spain topics like non-Euclidian projective geometry, number theory, and abstract algebra. Exploring current problems discussed in the most important centers of knowledge around Europe, García de Galdeano exposed Spain's conceptual backwardness and, at the same time, offered an opportunity for enlightenment. In sum, as he explains, the journal's readers would find "recompensa a sus nobles esfuerzos en la satisfacción que les produzca el haber contribuido a elevar el nivel científico de su patria, obra á la cual debemos concurrir todos los que nos interesamos por su prosperidad en la medida de nuestras fuerzas y en la esfera de acción en la que cada uno se mueve" (a reward for their noble efforts: the feeling of fulfillment for having contributed to increasing the country's scientific level, a task everyone interested in its prosperity must undertake to the extent their capacity and particular field of action allow).[37]

García de Galdeano's preoccupation with science education became even more pronounced when the journal reappeared in 1899 after a publication gap of more than four years.[38] The urgency with which the periodical was put back in circulation not only responded to the atmosphere of national pessimism that unfolded following the military defeat in the Spanish-American War, but also, as historian of science Mariano Hormigón explains, was a direct reaction to the government's plans to reform the university system.[39] The project led that year by Antonio García Alix, Ministro de instrucción pública (secretary of education), threatened most of the advances in science education that the country had achieved to date. In this context the author's marked interest in the German structures of teaching, researching, and applying sciences is of the utmost relevance.

While the educational structure in Germany, as noted earlier, was characterized by the proportional promotion of pure and applied sciences, García Alix's reform in Spain had reinforced an unbalanced allocation of resources that continued privileging practical over theoretical knowledge. García de Galdeano insisted on the importance of counteracting these changes and creating a new system of higher education that resembled the German framework. This proposal, however, presented many inconsistencies: not only was the German system internationalist and interdisciplinary by definition, relying on a national science already in place, but the very idea of adopting a foreign program interfered with the author's goal of strengthening the local scientific identity. Furthermore, the German model entailed the progressive development of scientific skills, favoring students' experiential learning first, to later focus on theoretical instruction. Although García de Galdeano agreed with this approach and encouraged its implementation in Spain, in his essays he had also emphasized that, in the formation of scientists, pure knowledge had to precede empirical understanding. These conflictual stances, nonetheless, were justified in the author's frantic search for avenues to expedite progress.

Similar contradictions would also complicate the journal's approach to the tension between science and religion. Within this context, the spiritual component of German philosophy of science, or Naturphilosophie—its understanding of nature as part of the Creation—became pertinent for García de Galdeano's social foundry. In the essay "La enseñanza de la ciencia matemática en la universidad," for example, he highlights the capacity of mathematics to mediate between the material world and the world of ideas—the realm of "los ideales del alma humana" (the human soul's ideals):

> Nos hallamos con la matemática, ciencia que parece reunir á las dos *Facultades* en un estrecho abrazo, pues mientras por un lado aspira a medir con sus cálculos y sus fórmulas los hechos ó los fenómenos de la Naturaleza, á esquematizarlos en sus fórmulas, á idealizarlos en sus leyes, por otro lado se constituye como un organismo *á priori* que radica en el fondo de las ideas y que, conforme con *nuestra* organización espiritual, se desenvuelve con fecundidad inagotable, con fuerza deductiva incesante y siempre creciente en un mundo más rico y más variado que el Universo entero.[40]
>
> (And we have mathematics, a science that seems to bring two *Abilities* together in a tight embrace: while, on the one hand, it aspires to measure facts and natural phenomena with its calculations and formulas, to schematize these facts in its formulas, to idealize them in laws; on the other hand, it constitutes an a priori organism with roots in deep ideas that, by agreeing with *our* spiritual organization, evolves with inexhaustible productivity, with endless and constantly increasing deductive strength, into a wealthier and more diverse world than the entire universe.)

By using mathematics as a constitutive part of the "spiritual organization" of society, science could finally find a resonance with national identity.

Because nineteenth-century Spanish print media functioned essentially as a political arena,[41] García de Galdeano's publication actively contributed to the institutional debate on education: *El Progreso* became an effective and persuasive channel of communication between the scientific realm and the political sphere. One of the main arguments put forward in García de Galdeano's social foundry is that scientific instruction entailed the progressive assimilation of concepts, and not only their memorization for practical uses. The notion that scientific knowledge needed to be useful to be worthy of instruction had precipitated a passionate debate among educators and politicians, leading to a partial paralysis at the state administrative levels regarding curricular or structural changes. The disparateness between theoretical and practical knowledge was already a point of contention in the educational reform advanced by Claudio Moyano in 1857.[42] Arguably the most important reconceptualization of public education until then, Moyano's articulation regulated the three components of the system—elementary, secondary, and university—by differentiating two stages in each level. Consequently, elementary school (*primera enseñanza*) was divided into basic and advanced; secondary (*segunda enseñanza*), into general and applied studies; and university (*nivel superior*), into departmental studies and technical or professional training. In all cases, concepts, theories, and ideas were taught mainly to support the production of practical applications. In his essays, García de Galdeano highlights some of the negative consequences of this utilitarian model: in Spain scientists served as technicians whose degrees and experience mostly addressed the country's shortages in terms of infrastructure and defense systems. Philosophical reflection and theorization were absent from higher education curricula, and in the case of mathematics, for example, only the most functional tools of calculation were taught at the university level.

Yet the importance of mathematics in higher education, both as a theoretical and a practical tool, was evident. Deeply influenced by German mathematician Felix Klein, García de Galdeano considered mathematics a main actor in the transformation of society and an indispensable element in the scientific cultivation of its members. If the country was going to be renovated and strengthened by scientific knowledge, it was necessary to have a clear definition of the role universities and other centers of instruction (circles of practical training, engineering schools, and so forth) played within a balanced academic system. In his general considerations at the beginning of "Estudios sobre la enseñanza," García de Galdeano exhorts the reader to reflect on the shortcomings of the educational system's current organization:

> No basta para nuestra cultura científica que nos queden como centros donde se refugie el culto de las ciencias matemáticas, físicas y químicas [en] las varias Academias y Escuelas de ingenieros, arquitectos, etc., puesto que en toda nación deben existir con absoluta independencia, como respondiendo á ideales diametralmente opuestos, la enseñanza científica de las Universidades y la de los Centros técnicos. . . . La Universidad se eleva á la idea, á lo abstracto, á la investigación que perfecciona y acrece el caudal de verdades para organizarlas en sistema. Las Escuelas y Academias técnicas se apoderan de todo lo más importante para las aplicaciones inmediatas. Aquélla aspira á lo fundamental, éstas á lo útil, y con arreglo á estas diferencias deben elevarse armónicamente en la nación los dos focos del saber científico.[43]

> (It is not enough for our scientific culture to have specialized departments in the many academies and schools of engineering, architecture, and so forth, where the devotion to mathematics, chemistry, or physics finds a shelter, because in every nation there must exist, with absolute independence, like

following opposite ideologies, scientific instruction in universities and scientific preparation in technical centers. . . . Universities have to deal with the ideal, with abstraction, with research that increases the number of truths and organizes them into a system. Technical schools and academies take control of the most important elements that immediate applications demand. The former aims at the fundamental, the latter at the useful, and with such differences these two centers of scientific knowledge should harmonically coexist and rise in the country.)

For the author, research and instruction were tasks that needed to be carried out by different academic entities. Without contradicting his plea for an integrated system of education in which theoretical and applied sciences complemented each other, this suggestion would allow mathematics to be considered more than an instrument for applying abstract knowledge, namely, to be seen as an expression of the fundamental principles of nature. Both universities and institutes could thus use mathematics in different but indispensable ways: while universities would privilege research and formation, institutes would promote material accomplishments and factual transformations achieved through the application of knowledge.

This compartmentalization, however, was never effectual in Spain, as García de Galdeano explains in his essay "La enseñanza de la ciencia matemática en la Universidad." In part, this situation was due to the concentration of social and political efforts, not in strengthening the country's institutional body, but in defining and defending national identity:

> Basta para explicar y disculpar el hecho excepcional que ofrece España, . . . de su aislamiento científico, el recordar la vida toda de nuestra nación que constituye su historia, y que se resume en una guerra de raza sostenida con perseverancia incansable por espacio de ocho siglos que ha concentrado allí todo el esfuerzo patrio, y que no le ha permitido distraerse hacia otros fines.[44]
>
> (To explain and excuse the exceptional situation of Spain, . . . and its scientific isolation, it would be enough to remember the life and history of our nation, one that may be summarized in a racial war upheld with tireless perseverance for eight centuries, on which all the patriotic efforts were concentrated, preventing the country from pursuing different goals.)

While other European countries focused their energy on devising systems for materializing their modernization by means of promoting science, through both research and the practical application of knowledge, Spain concentrated its scientific efforts on military campaigns and plans of territorial expansion. Under those conditions, science education was used primarily to supply the tools for such endeavors, and research rarely enjoyed institutional support. The ideological divergence between those educators and administrators who understood science as philosophical inquiry and those who mainly saw it as a practical instrument would reach its peak when the discussion turned to mathematics, a science generally considered useless if studied in its pure form. In this regard, García de Galdeano agreed with Marcelino Menéndez Pelayo in asserting that the development of the country would not be possible until society was able to see "la sublime utilidad de la ciencia inútil" (the sublime usefulness of the useless science).[45]

Criticism of the institutional emphasis on the practical uses of science can be traced back to the first issue of the journal. Published in 1891, the opening editorial article, "El objeto y los propósitos de la actual publicación" (The object and goals of the present publication), makes an assessment of the obstacles and opportunities of the local educational system by exalting recent advances in mathematics produced in other European countries. After describing the work of Arthur Cayley, James Joseph Sylvester, and William Hamilton, García de Galdeano adds,

> Todo este movimiento general se sigue en España, si bien reducido á aquello de más esencial que contiene cuanto ofrece inmediata aplicación, ó que es un elemento indispensable para llegar al nivel de lo que hoy constituye la cultura matemática; pero esto que se circunscribe á un reducido número de personas dedicadas por su profesión á tal género de lucubraciones, no transciende a otro público más numeroso que podría aprovechar las ventajas de estos conocimientos, y aun llegar á contribuir al adelantamiento general, si se le facilitara el acceso a ideas que parecen más difíciles de adquirir al parecer de lo que son en realidad. Estas dificultades aparentes son consecuencia de ignorarse los fundamentos de las varias teorías ó ramas que constituyen las ciencias matemáticas.[46]
>
> (All these advances are studied in Spain, although in a limited form wherein contents are reduced to those that offer immediate application or use basic elements of everything that mathematics provides nowadays; but these ideas, circumscribed to a reduced circle of people whose professions entail this kind of reflection, never transcend to a larger audience who might take advantage of this knowledge and even contribute to the general advancement of society if they had access to ideas that seem more difficult than they actually are. These apparent difficulties are just the result of not understanding the foundations of the different branches and theories that constitute the mathematical sciences.)

Considering that pure knowledge, or theory, always preceded applicable knowledge, here the author provides the foundational ideology for his projected system of higher education—a structure that, through different levels of instruction, would allow the reconciliation of the "estudios positivos de aplicación" (positive applied studies) and the moral strength of what was known "desde antiguo con el nombre de humanidades" (for a long time as the humanities).[47]

García de Galdeano recognized education to be moving between opposite forces: theory and application, philosophy and science, mathematics and engineering. For him, the German system of instruction, where students' preparation for research at the university level started early in secondary school (*realschule*), was a successful example of how Spain's pedagogical reform should be executed. An early emphasis on science would change pupils' expectations and achievements once they were preparing for college in specific professional high schools (*gymnasiums*). To support this effective structure, a network of "material mediums to favor culture" had to be in place. These included, among others, state support for the creation, maintenance, and improvement of libraries, museums, and other similar venues, real havens for the objects to which they were devoted.[48] A configuration of this sort would be the most effective remedy for the lack of a strong scientific tradition; its implementation would make Spanish education comparable to that of the most advanced European countries.

Interestingly, this reformulation also acknowledged the government's demands for practical instruction to contribute to immediate needs. Similar to Vicuña's social foundry, García de Galdeano considered feasible the reorganization of the system already in place, instituting academic spaces where pure sciences could be developed in the same measure as applied knowledge. This appropriation of the German model, in any case, was somewhat modest—inspired by the Krausist search for a harmonic balance of spiritual and material forces, his plan mainly required a change of pedagogical perspectives. Within this configuration, therefore, *institutos* (institutes) and *academias* (academies) were to concentrate on the instruction of technical skills, and *universidades* (universities) were to focus on research. Should the state follow these recommendations, the country would be finally able to materialize its own national science:

> [Es necesario] crear una escuela científica nacional con iniciativas, con impulso propio, no como reflejo ó simple imitación, como producto exótico importado de otras regiones, sino como fruto de la meditación propia, de la actividad moviéndose libremente para crear en la región de las ideas, ó para expresar con estilo propio aquello que la inteligencia elaboró y asimiló á su esencia, para presentarlo con el sello de la personalidad.[49]
>
> ([It is essential] to create a national scientific school with its own initiative and impulse, not as reflection or simple imitation, or as an exotic product that is imported from other regions, but as the result of local meditation, of ideas moving freely to express in its own style what intelligence elaborated and assimilated as part of the local essence, and to present it with the mark of its own personality.)

Notwithstanding the evident inspiration in foreign prototypes (particularly the German one), in the proposed system there had to be an element of originality that reflected the national personality. To be sure, García de Galdeano's social foundry establishes a parallel between the system of scientific scholarship and the development of a modern nation. For him, an early introduction to the scientific world would prepare young students for the metaphysical and philosophical questions they would need to ask later, during their doctoral studies. Likewise, the nation as a whole had to be educated so its constituencies could address more elevated matters while still generating the supplies for immediate material needs.

Since the future of the nation depended on the development of practical applications to generate a solid economic ground, García de Galdeano's plans gave special consideration to technical instruction. In his view, centers dedicated to this kind of training were fundamental to producing the human resources to work on the basic needs of the country. He hoped Spaniards could somehow profit from their tendency to privilege the practical over the theoretical. Recognizing this possibility, however, did not imply that the country was going to be unable to develop its own science at some point. In fact, as the author promptly recognizes in one of his essays, it was not the case that Spain lacked the talent: "no han faltado los destellos del genio ni los esplendores de épocas de grandeza para las ciencias y las letras" (The brightness of geniuses is not absent, nor the moments of splendor in sciences and letters).[50] The problem was that these sparks of national brilliance had always been extinguished by the restrictions of a misconceived educational apparatus and the burden of tradition.

Given Spain's exceptional condition, García de Galdeano believed that the simple adoption of a foreign model would not suffice to solve all problems. In Spain, the notion of education as a balanced place for the production of practical results and the conception of theoretical ideas had to overcome more complex and unresolved ideological tensions between progressivist and traditionalist political agendas. His suggestions to find an adequate and productive equilibrium between these two perspectives, nonetheless, would resonate with recommendations offered by other scientists and educators.[51] Ultimately, the main advantage of the structure put forward for consideration in García de Galdeano's social foundry was that the interest in science, both as tool and as cognition, could be revitalized, allowing the rise of a national scientific culture that profited from the peculiarities of the Spanish situation. Thus, the vicious circle that related the lack of scientific development to the wrongful prioritization of practical education, and the absence of practical solutions to the deficiencies in science education, would finally be broken.

Focusing on the spiritual aspects of the scientific mission, other scientists of the period proposed strategies for constructively aligning education and national development. Such was the case of chemist José Rodríguez Mourelo. Condensing most of the ideas discussed in this chapter, in the closing section I show how his social foundry endorses the search for a local, national scientific tradition chiefly rooted in moral principles.

José Rodriguez Mourelo and Education's Capacity for Moral Transformation

In 1903, José Rodríguez Mourelo was inducted into the Royal Academy of Sciences after a long and successful career as a professor of chemistry in Madrid, where he also worked for newspapers and magazines such as *El Heraldo* or *La España Moderna* (The modern nation). Proof of his commitment to education and the promotion of science is his long record of publications and public talks, in which he addressed both general and specialized audiences who were interested in the study and application of chemistry. In his acceptance speech at the Royal Academy of Sciences, for example, he discussed what he saw as the obligations of every Spanish citizen with regard to the moral, material, and social transformations of progress. Rodríguez Mourelo questioned the emphasis on the practical, applied, and consequently shallow uses of knowledge in a country that still needed to acknowledge the spiritual importance of science. Although some of these ideas were also at the center of Vicuña's social foundry discussed earlier, the moral component of scientific practice was not in consideration—here, instead, science is depicted as one of the missing or defective building blocks of the national character.

Following a very similar professional trajectory to that of Laureano Calderón, Rodríguez Mourelo served as an instructor at the Institución Libre de Enseñanza, a position he held until he was appointed, in 1900, to the Escuela de Artes e Industrias de Madrid (School of Arts and Crafts of Madrid), where he taught organic and inorganic chemistry. He also shared Calderón's support of Darwin's and Spencer's ideas, as well as many of his propositions regarding higher education. Very active within his academic circle, Rodríguez Mourelo published important studies on fluorescent materials, gaining local and international recognition.[52] Besides his work as a researcher and teacher, one of his major contributions to national science was through his labor as a public speaker and promoter of educational reform; in these capacities, he demonstrated great versatility in approaching the country's problems from a scientific perspective. His particular way of looking at these challenges was equally imbued with his vision as educator, hence the recurrent revision of the issue of scientific instruction in most of his reflections.

For Rodríguez Mourelo, Spaniards' unwillingness to experience the world was one of the biggest barriers to modernization. Considering that science needed to be supported by faith and devotion, in his social foundry nature takes the place of God—the cult of science thus consisted of understanding and deciphering the mysteries of nature, a task for a new generation of thinkers and enthusiasts, and a regenerative exercise that could finally break the country's attachments to the past. In his talk at the Academy of Sciences, he explained,

> Fe en la ciencia, voluntad para educarnos en ella, aprovechando sus principios redentores, amor al trabajo, amor á la naturaleza, amor á a humanidad, es lo que necesitamos predicar sin tregua en todas partes, para reconstruir el alma nacional, hoy sin voluntad y como nunca necesitada de juveniles energías; rayos de sol, atmósfera pura que le infundan nueva vida; vida de realidad, conforme es la vida de la ciencia; vida de alegría, sin misterios ni sombras, cada vez más perfecta, más grande y más vigorosa.[53]
>
> (Faith in science, intention to educate ourselves in it to take advantage of its redeeming principles, love for work, love for nature, and love for humanity; those are the values we need to promote everywhere without rest, so that we can rebuild the national soul, empty of will today and more than ever in need of young energy; in need of sunlight and a pure atmosphere that fills it with new life; a real life, as the life of science should be; a life of happiness with no mysteries or shadows, each time more perfect, bigger, and stronger.)

By associating nature with a clean source of power and energy, Rodríguez Mourelo opposes a dark past of technological stagnation to a positive, radiant, and bright present founded on a solid education and an active agenda of scientific research. By supplying the country's engine—which he calls the "national soul"—with the redeeming power of science, it was possible to generate moral and material conditions of prosperity and development.

The creation of sustainable improvements in the country, nonetheless, required a socially based approach in which scientific research and technological development operated as active agents of spiritual progress. Pointing to the transformative, liberating capacity of science throughout history, Rodríguez Mourelo added, "Como educadora de la inteligencia, de la voluntad y del sentimiento y en su calidad de emancipadora, tiene la ciencia muy elevados fines sociales y de tal suerte los cumple, que en nuestro días han llegado á transformar, por los inventos de todo género, por la instrucción, por las investigaciones, y más que nada por la educación, el modo de ser de las sociedades humanas" (As the educator of intelligence, of will, and of emotions, and in its emancipatory capacity, science has very elevated social purposes, and it fulfills them so well that in our days these goals, through all sorts of inventions, instruction, research, and more than anything, through education, have produced the transformation of human society).[54] If the main role of education was being an instrument of science, research and innovation depended on an adequate structure of the educational system. In this scenario, the consolidation of a scientific culture was inevitably linked to the transformative capacity of society. Such potential for change, however, could not emerge spontaneously; it required society's will to work toward a common goal.

For Rodríguez Mourelo there were two Spains: one with a fear of social transformation that had diverted the progress of science and delayed industrialization, and one that had strived to embrace modernization in spite of the profound impact that its social changes could have on national identity. Though Spain did, strictly speaking, have a scientific tradition, that culture had been dissociated and even alienated from identity. In this context, the consolidation of an adequate educational system would require a double effort:

> Habiendo perdido, para nuestro mal, la herencia, olvidando el pensamiento científico español y apartándolo de los naturales desarrollos que sin duda hubiera tenido, es menester realizar un doble trabajo que comprende estos dos extremos: reconstruir lo tradicional, aprovechando para ello tan solo lo verdadero y prácticamente útil, y adaptar al medio nacional los bien elegidos elementos externos destinados a formarlo; ésta es la primera labor educativa de la ciencia, construir el medio en el cual ha de desenvolverse, formar el ambiente de perfección, crear la necesidad de educarse, como único instrumento de vida.[55]
>
> (Having lost, to our disgrace, the tradition of science, forgetting that there was a Spanish scientific thinking, and distancing it from the normal developments that, without a doubt, it would have produced, it is necessary to execute a double task that encompasses these two extremes: rebuilding tradition, only taking advantage of that which is original and practically useful; and adapting to the national construction only those external elements that were originally conceived to form part of it. Such is the first educational goal of science: to build the environment in which it can evolve, to produce that atmosphere of perfection, to generate the need to consider education as the main instrument of life.)

Any plan of educational reform, therefore, first needed to facilitate a medium wherein science and technology could emerge naturally as products of national ingenuity. Yet importing and adapting foreign educational models and research strategies were also necessary tasks in the process of forging a national science. Scientific tradition would then entail a contradictory combination of local and foreign elements whose only condition for being adopted within the national program was that of being practical, thus adding another layer to the debate on the utilitarian purpose of science.

In his social foundry, Rodríguez Mourelo resorts to elemental notions of energy, work, and power to emphasize the importance of finding balance between practical and theoretical knowledge and between local and foreign ideas in the conception of a modern educational system. While consolidating Spain's national science implied the defense of identity, the appropriation of foreign paradigms to model local academic practices was inevitable—only by imitating successful cases would Spain be able to reach its own paradigm for modernization. However, the country continued privileging the practical uses of science and its immediate applications over an in-depth study of principles, concepts, and theories. In his response to Rodríguez Mourelo's inaugural speech at the Royal Academy of Sciences, Echegaray asked, "¿Qué debe proteger más una nación, la alta Ciencia, la Ciencia pura, ó la Ciencia práctica y de aplicación inmediata á la Industria?" (What should a nation protect more: the high Science, the pure Science, or the practical Science with immediate applications to industry?).[56] And by "protect," Echegaray implied the promotion and support of institutions that guaranteed both the production of knowledge and the development of skills to supply the country's immediate needs: roads, factories, military defense, and infrastructure.

The answer to Echegaray's question, however, was not a simple one. As he would later recognize in the same speech, practical science was intrinsically connected to theoretical developments, which, in turn, could not be achieved without the capital and resources produced by technology and other applications of knowledge. The pattern of these mutual dependencies could be compared to the water cycle, wherein clouds produce the rain that feeds the rivers, which, in turn, provide sustenance for the clouds—there would be no rivers without clouds, but no clouds are possible without rivers: "Pues una relación análoga tiene la Ciencia pura con las ciencias de aplicación: faltando aquellas se esterilizan éstas" (Pure science hence has an analogous relation with the applied sciences: if the former is not present, the latter becomes unproductive).[57] In contrast to Vicuña and García de Galdeano, whose social foundries suggest specific strategies to improve the system of higher education in Spain, Rodriguez Mourelo's and Echegaray's reflections fall short in advancing concrete plans to balance practical and theoretical knowledge and secure the local production of science.

As I have shown in this chapter, social foundries based on the analysis of the educational apparatus tended to align with the romanticism embedded in the Krausist notion of national spirit. This is not a coincidence if one considers that the conception of the nation as a product of certain geographic conditions, connections to the past, and reverence for tradition was first developed by German philosophers—in his appropriation of Krause's doctrines, Sanz del Río also imported the ideas of Johann Herder.[58] Under these premises, the Spanish spirit was ill, and one of the diseases afflicting it was the lack of an adequate system of education. Thinking about the country in terms of a sick body functioned as an alternative space for social criticism, within which it was possible to diagnose with precision and authority the impediments to the consolidation of the modern nation.[59] Chapter 4 explores other avenues of social diagnosis informed by industrial and scientific development and articulated through literature. These reflections expose the heated social tensions that arose from the growth of the proletariat, and delve into problems such as class disparity, labor exploitation, and accumulation of capital in relation to identity.

CHAPTER 4

Social Engineering

In contrast to evaluations of the country inspired by economic, scientific, or educational projections, this chapter studies assessments built on the social tensions that arose from the adoption of specific economies of labor in *fin-de-siglo* Spain. The conflictive transition from systems anchored in feudal practices to industrial apparatuses of production based on a capitalist structure brought into question the role of regional and state nationalisms, as well as the influence of religion in the consolidation of certain ideological agendas. Whereas previous approaches highlight the incompatibility between industrial modernization and national identity, the diagnoses here focus on the social repercussions of progress. I analyze works centered on the mining and steel industries by authors Vicente Blasco Ibáñez, Manuel Ciges Aparicio, and Concha Espina, and review their context of production. The *social foundries* proposed in these texts look at the role that different social actors—the Church, the aristocracy, industrialists, workers, activists, and so forth—played within the industrial milieu and evaluate the way in which they assimilated the new realities of modernization. While Blasco Ibáñez's novel *El intruso* (The intruder) (1904) shows the confrontation between workers and tycoons under the vigilant eye of the Church, Ciges Aparicio's chronicles *Los vencedores* (The defeaters) (1908) and *Los vencidos* (The defeated) (1910) contrast two accounts of a major strike organized and executed in response to unfair labor practices that in many cases were endorsed by the Church and the local industrial bourgeoisie. Finally, Espina's novel *El metal de los muertos* (*The Metal of the Dead*) (1920) focuses on religion and its involvement in the labor question, problematizing the Church's double part as moral support of the exploited and economic ally of the exploiters.

Anticlericalism and the Social Question in Vicente Blasco Ibáñez's *El intruso*

Vicente Blasco Ibáñez's participation in the public sphere was marked by his involvement in national politics, his work as a journalist, and his international success as a writer. Some of his most prominent novels were adapted into films in the United States, where he gained fame, fortune, and recognition. As a politician, Blasco was a staunch opponent of the political system of the Restoration. During his career, he represented different republican coalitions and openly criticized the ideological foundations and implementation of Cánovas's plans. However, and despite his influential personality and profound commitment to liberal ideas, Blasco's participation in government was scant—his contribution to national politics was limited to his long tenure as a member of the parliament between 1898 and 1907. As biographer Ramiro Reig explains, this lack of involvement in more active and visible positions responded to his own conception of a modern state—a vision opposed to that of the prevailing Canovismo.[1] Blasco was a supporter of a reformist agenda that sought, among other progressivist changes, the effective separation of the Church and the state, agrarian reform, equalitarian access to education, and universal suffrage. He also believed in the need for urban renovation programs designed to accommodate cities and towns to the particular demands of industrialization, and became involved in the development of plans for the expansion of his native city, Valencia. As a journalist, Blasco founded the newspaper *El pueblo* (The people), and used this space to promote the need for social reform in the country. He was also a regular contributor to the most prominent newspapers and journals of the period, where he published influential editorial pieces on current affairs in Spain and the rest of Europe.[2] In the literary field, his plurality of interests, the tendency to cultivate different genres in his writing, his innovative spirit, and his active production made him one of the most respected and successful writers of his time.[3]

Blasco's production is vast, and some of his works have gained more recognition than others. Critics have paid more attention to texts that are, to a certain extent, representative of the author's political ideology or of his particular artistic approach.[4] In contrast to their analyses, in this section I focus on one of Blasco's less studied works, *El intruso*. Published in 1904, the novel addresses industrial modernization and discusses the economic and cultural shock that its social transformations produced on an otherwise conservative society—that is, with a rigid social structure and profound religious convictions. I propose that, in contrast to other contemporary novels studied in this book (see, for example, the analysis of Palacio Valdés's novel *La aldea perdida* in chapter 6), Blasco's social foundry not only criticizes the conditions under which industrial development was possible in Spain, but also resorts to regional nationalism to depict the conflictive appropriation of the logics of modernization in an ideologically fragmented society.

While in the novel there is a clear acknowledgment of the role that the past and tradition played in the articulation of national identity, at the same time there is an emphasis on the financial and social possibilities that industry's economic and cultural changes offered. Blasco's social foundry shows the problematic assimilation of the mechanisms through which industrial modernization (re)shaped national identity. To illustrate this, the novel develops multiple storylines that converge in the recount of a violent confrontation of two factions with different, even opposite, ideas about progress: religious nationalists and business aristocrats, on the one hand, and immigrant workers and the progressive bourgeoisie, on the other. Although critics have long been aware of Blasco's cycle of *novelas sociales* (novels with a focus on social issues, which, in addition to the text analyzed here, include *La catedral* [The cathedral] [1903], *La bodega* [The fruit of the vine] [1904–1905], and *La horda* [The horde] [1905]), few have looked in detail at the importance of *El intruso* as a testimony of the moral implosion of traditional society and the conflictive realignment of economic forces around industry.[5]

Structured around the story of two main characters, the wealthy steel tycoon José Sánchez Morueta—a representative of the conservative aristocracy—and his cousin, the physician Luís Aresti—a progressivist sympathizer with the workers' causes—the novel's plot attests to the social and political dichotomy projected onto their circumstances. Readers then witness their progressive ideological transformation from moderate positions, toward religious fanaticism in the case of Sánchez Morueta and toward political radicalization in the case of Aresti. These escalating transitions also expose the clash of two temporalities—two different ways of understanding the world and assimilating social change. A good example in the novel of this articulation is the way in which the industrial city is depicted as a space of passage from the passivity of the rural past to the dynamism of the industrial present:

> La antigua Bilbao de los comerciantes y los marinos, que aun no conocía el valor del hierro, era más feliz con la paz de un trabajo lento y ordenado y la llaneza fraternal de sus costumbres, que la villa moderna, con sus improvisadas fortunas, sus ostentaciones locas y aquella riqueza disparatada y rápida que apenas si dejaba en el país rastros beneficiosos de su paso, perdiéndose en las obscuras tragaderas del intruso negro, aparecido en la hora suprema de la fortuna para sentarse al lado de los favoritos de la suerte, ofreciéndoles el cielo á cambio de una participación en el botín.[6]

> (The old Bilbao of traders and sailors, that city that did not know yet the value of iron, was happier with the peace of a slow and ordered labor, and with the fraternal simplicity of its customs, than the modern village with its improvised

> fortunes, crazy ostentation, and ludicrous and fast wealth that did not produce tangible benefits to the country after circulating, wasted in the dark drain of the black intruder, who arrived in Fortune's best hour to sit on the side of those favored by Luck, offering them Heaven in exchange for a portion of the booty.)

Whereas the rural world was "slow and ordered," the "modern village" is frenetic and chaotic; one guaranteed its inhabitants' well-being, the other is ruthless and cruel. Blasco's description of the industrial economy shows his concerns with the fact that the main beneficiaries of industrial progress were not the workers but the same social sectors that had controlled the wealth in the past. Modernization's ability to transform society had been somehow distorted, and the working classes were victims of a new order wherein their living conditions were worsening. As a matter of fact, the proletariat could not "get a portion of the booty" that modernization generated, a reward that was instead diverted to the Church's treasury, the "black intruder" mentioned in the passage and the inspiration for the novel's title.

Through contrasts between past and present, the novel also highlights the physical and moral effects of industry. Interestingly, in these descriptions the countryside is not presented just as an idyllic space—a tendency in writers who saw the past with a certain nostalgia—but as a site of destruction. By displaying the multiple transformations of nature as the result of industrial practices, rural spaces appear as crumbling ruins of the past. At the same time, modernization is staged as a dynamic yet damaging force, a dialectic exemplified in Aresti's description of the outskirts of Bilbao:

> Por un lado del tren se abarcaba el vertiginoso movimiento de la ría con sus barcos y fábricas; por la ventanilla opuesta admirábase la paz de los campos, el trabajo cachazudo y tranquilo de los aldeanos removiendo la tierra arcillosa. Las mujeres, con la falda atrás y las piernas desnudas, sudaban dobladas sobre el surco. Las vacas movían el baboso hocico, sin ninguna inquietud, al ver el tren y volvían de nuevo á rumiar con la cabeza baja sobre el verde del prado. Grupos de mujeres lavaban sus guiñapos casi tendidas al borde de arroyos de líquido rojo, como si fuese sangre. Era el eterno color del agua en los alrededores de Bilbao: los lavados del mineral enrojecían hasta la corriente del Nervión. La industria, al enriquecer al país, corrompía las aguas puras y cristalinas de la época pastoril.[7]

> (On one side of the train, one can see the vertiginous movement of the river with its boats and factories; through the opposite window, one can admire the peace of the lands, the slow and peaceful labor of the villagers removing the clay soil. Bent over the furrow and sweating, there are women with rolled up skirts and naked legs. On the green meadow, a herd of cows move their slobbery snouts and return to chew the grass without any concern after seeing

> the train. There are groups of women washing their tatters, almost laying on the edge of the red streams of a liquid similar to blood: it is washed mineral from the mines that even reddens the currents of the River Nervión. Industry, while enriching the country, spoiled the pure and crystalline water of the rural times.)

The association between water and blood functions here as a referential marker of a reality defined by industry's disruptive capacity. The "vertiginous" movement of industrial machinery had displaced the "peace" and "purity" of rural spaces to replace them with noise and pollution. This image also apposes manual labor to mechanical work, emphasizing the role industry played in potentializing workers' efforts as essential contributions to the country's future. In the border that separates these realities, rural and industrial, there was the train—vehicle and image of progress that prods Aresti's precise diagnosis of society's afflictions. In fact, it is the liminality generated by the railroad that prompts his vision of the two historical dimensions of national identity.

When Blasco visited Bilbao at the outset of the century to do research for his novel, he was deeply impressed with the complexity and size of the industrial economy. Machines had replaced animals as working tools, and the noisy movement of the intricate and large apparatuses of production disrupted that which once had been a peaceful agrarian society.[8] Descriptions of Bilbao's metropolitan area in the novel attest to this reality:

> Y flotando por encima del bosque de chimeneas de ladrillo y de hierro, el eterno dosel de la moderna Bilbao, los velos en que se envuelve como si quisiera ocultar púdicamente su grandeza, los humos multicolores de sus fábricas, negros, de espesos vellones, como rebaños de la noche; blancos, ligeramente dorados por la luz del sol; azules y tenues como la respiración de un hogar campesino; amarillos rabiosos con un chisporroteo de escorias minerales.[9]

> (And floating on top of a woods made of brick and iron chimneys—modern Bilbao's eternal facade—one can see the veils with which the city wraps itself as though it would modestly be seeking to hide its greatness or the multicolored smoke of its factories: black, with thick fleeces as flocks of sheep in the night; white, lightly gilded by the sun light; blue and weak as smolder in a peasant's hearth; extreme yellow with sparkly dross.)

By showing how factories, with their chimneys and smoke, had replaced flocks of sheep, woods, rivers, and peasants' hearths, this passage subtly confronts past and present, evincing a shift in the perception of reality that Blasco incorporates into his social foundry.

The use of the human body as an allegory to describe the industrial apparatus complements Blasco's diagnosis of the social ailments. Detailed descriptions of different stages in the process of mineral extraction and metal production as circulatory, muscular, or digestive systems, for example, highlight the organic nature of the national geography and its environmental frailty: "Aresti, ante este desgarrón de la corteza terrestre que mostraba al aire sus entrañas, reordenaba las formas y colores de las piezas anatómicas reproducidas en sus libros de estudio. Las calizas blanqueaban como huesos; las fajas de mena rojiza tenían el tono sanguinolento de los músculos, y las manchas de tierra vegetal eran del mismo verde musgoso de los intestinos" (Aresti, beholding the big tear on the Earth's crust that exposed the entrails of the planet, reordered in his mind the shapes and colors of the anatomical parts his medical textbooks reproduced. The limestone whitened like bones, the frames of reddish ore had the same blood-like tone of muscles, and the stains of vegetable soil were similar to the mossy green of intestines).[10] Descriptions of the landscape's different tones and shades operate here not only metaphorically, through images, but also poetically, through discourse—the novel thus proposes the study of society as both scientific and literary efforts.

This use of poetic figures in the description of industrial processes also confers a mythical component on the allegorical representations of society. In contrast to classical parables wherein references to mythology characterize actors and situations, in *El intruso* these associations operate on a more abstract plane. Blasco traces symbolic links between the production of metals, the primeval nature of fire, and different images of destruction and suffering: "Mientras el ingeniero detallaba sus explicaciones, el médico, asombrado por la enorme mole de las dos torres ardientes que parecían servir de pilares al firmamento, pensaba en el culto del fuego, en la adoración de las razas antiguas al gran elemento creador y destructor, en los ídolos ígneos que cocían dentro de su vientre, en repugnante holocausto, las víctimas humanas" (While the engineer expanded on his explanation, the doctor, amazed by the enormity of the two scalding towers that could be pillars of the firmament, was reflecting on the cult of fire, on the adoration of ancient races for the great element of creation and destruction, on the igneous idols who, in a revolting holocaust, cooked their victims in their entrails).[11] Industry appears here not only as a space of creation and destruction linked to the origins of the world but also as one of the "pillars" of contemporary society; it *brings light* to progress thanks to the workers' labor. The image of an industrial machine that devours workers or incorporates them as cogs in its complex mechanisms gives an atemporal and, at the same time, sacred nature to modernization, thus allowing the cognitive negotiation of its destructive or generative power. It is in this regard that allusions in the text to Vulcan's forge or Prometheus's theft underline the high cost of progress.

Poetic language reaches its highest point in the narrator's depiction of steel inception, an image with religious undertones that generates new elements of contrast for evaluating the social tensions between workers and industrialists:

> Se abrió un pequeño agujero en la base de una de las torres y apareció un punto de luz deslumbradora, una estrella roja de agudos rayos que herían la vista. Se fue agrandando, y un arroyo rojo obscuro, como de sangre de toro, corrió por la tierra con un chisporroteo ruidoso. . . . Aresti admiró a los trabajadores, que estaban allí como en su casa, habituados á una temperatura asfixiante, moviéndose como salamandras entre arroyos de fuego, enjutos, ennegrecidos cual momias, como si el incendio hubiese absorbido sus músculos; dejándoles el esqueleto y la piel. . . . No era una estrella lo que se abría en la tierra refractaria: era una gran *hostia* de fuego, un sol de color de cereza, con ondulaciones verdes, que abrasaba los ojos hasta cegarlos.[12]
>
> (A small opening was created on the base of one of the towers, and a minuscule point of blinding light emerged, a red star of sharp beams that hurt the eyes. Once it grew, a dark red river, as if it was made of bull's blood, ran on the ground with a noisy frizzle. . . . Aresti admired the workers, who acted as if they were at home: accustomed to the asphyxiating temperature, moving like salamanders around rivers of fire, scrawny, darkened as mummies—as if fire had absorbed their muscles leaving only bones and skin. . . . It was not a star that was being born out of the refractive ground; it was a big fire *Host*, a sun with the color of cherries and green curls that burned the eyes until it blinded them.)

Steel, in this description, is a bright star that concentrates the energy of civilization. The cost of producing that *light*, however, falls back onto the workers. Authors analyzed in previous chapters problematize this social setting in a similar way: for Echegaray, Mallada, or Calderón, for example, the working class needed to join efforts in order to produce the transformative energy of progress. Here, workers are components, tools, and nourishment for the industrial machinery—they go from being "salamanders" to "darkened mummies" once their muscular system is extracted and absorbed by the machine as fuel for the production of wealth. Within this exploitative dynamic, industrial development is depicted as a negative force that victimized the working class.

Even though Blasco recognized the important role that the bourgeoisie could play in addressing the *labor question*, he felt skeptical about the extent to which there was a real commitment on its part to effect social changes. In fact, *El intruso* denounces the reluctance of the middle and professional classes to actively participate in creating the conditions for an egalitarian society, an aspect evinced by their interest in accumulating wealth and participating in speculative finances.[13] Yet apathy was only part of the problem—the Church and the aristocracy also contributed to this misappropriation of the social principles of modernization by instigating a class segregation based on nationalistic feelings, which, in turn, gave them wider control over the economy. In Blasco's social foundry, society is codified through multiple contrasts: religion and capitalism, tradition and modernity, reason and faith. In the negotiation of these opposites, violence is symbolized as industrial contamination, a destructive force that prevented the coexistence of traditionalist and progressivist ideas. The novel thus targets the political system that allowed the persistence of old economic practices at the cost of the common good.[14]

As historian Miguel Artola explains, the multiple ideological tensions that characterized nineteenth-century society were reflections of the incompatibility between traditional economic structures and the emergence of social conditions stimulated by industrial growth.[15] One of the most tangible consequences of this struggle was the translation of feudal exploitative practices into the realm of industry, a situation that created an environment of unconformity and unrest among workers. As pointed out earlier, many of these practices were scientifically justified and promoted as indispensable for the functioning of society. The use of thermodynamics to defend class segregation or the implementation of labor-intensive models, for example, was supported on the idea that the unequal distribution of wealth was necessary for the adequate circulation of capital and work. Along with science, religion was also used as a tool of ideological manipulation that gave grounds for exploitation and social stratification.

Whether it was supported by scientific notions or religious precepts, political ideology operated as a flexible system of values that could be tailored to fit the specific needs of the group in control of the industrial apparatus. This latitude is illustrated in the novel through the progressive transformation of the main characters. Steel tycoon Sánchez Morueta, for instance, transitions from being a prototype of the progressivist, entrepreneurial bourgeoisie to a puppet of the Church; Aresti, in turn, moves from sympathizer with the causes of the proletariat to active participant in its social revolt. The growing friction between the two characters throughout the story, which turns a simple family quarrel into a physical brawl at the end of the novel, reiterates the image of two contrasting realities—the peaceful past and the frenetic present. The characters' personalities and complexion also illustrate this tension. Sánchez Morueta, for example, is first depicted as a strong and energetic character: "Tenía cana la barba, gris el pelo y, sin embargo, parecía envolverle un nimbo de juventud, de fuerza serena, de energía reposada y tenaz, que se comunicaba a cuantos le rodeaban . . . [luchaba] sin más auxilio que las energías del músculo y del pensamiento, y acababa por posesionarse del mundo" (He had a gray-haired beard, gray hair and, nonetheless, it was as if a halo of youth was surrounding him, an aureole of sober strength, of rested and tenacious energy, that easily involved people around him . . . [he fought] with no other aid than the energy of his muscles and reasoning, and he ended up owning the world).[16] But, later, when the businessman has already succumbed to the dark forces of the Church, the narrator presents him as a timid and spiritless man unable to hold the reins of his own emporium: "Realmente, el grande hombre no gozaba de buena salud. Había adelgazado mucho, su barba era casi blanca, lo ojos los tenía hundidos, y en su rostro enjuto se marcaban los pómulos con agudas aristas, pareciendo la nariz más grande y pesada" (To be honest, the great man was not in good shape. He had lost a lot of weight, his beard was almost white, his eyes were sunken, and in his scrawny face the cheekbones were marked by sharp edges, which made his nose look big and heavy).[17] Frailty, as showed here, was not only a manifestation of his physical condition but a symptom of his losing control of the productive machine.

In contrast to the weakening of Sánchez Morueta, Aresti's affirmation of *faith* followed a different path. The physician is depicted as a strong believer in social transformation, a process that in his view was not only possible but necessary. While the spiritless businessman was distracted by abstract ideas about sin and death, the doctor understood the concrete value of work, a force that religion could not diminish: "El hombre moderno no debía perder el tiempo preguntándose sobre el origen del mal ó si la naturaleza está corrompida por el pecado: las dos grandes preocupaciones de la moral cristiana. *Bastábale saber que la naturaleza, buena ó mala, se modifica ó transforma por el trabajo*" (Modern men did not need to waste time questioning the origin of evil or if nature was corrupted by sin: the two main concerns of Christian morals. *It was enough to know that nature, good or evil, is transformed or modified by the action of work*).[18] Insofar as it allowed the transformation of nature to produce economic profits, work was "above good or evil." In Arsesti's mind, ideas did not need to become dogmatic; on the contrary, they were the fuel of society's remodeling and transformation: "No: yo no soy liberal: yo soy un hombre de mi tiempo, tal como me han formado las circunstancias de mi país, y no como lo enseñan los libros. Yo soy un jacobino; un hombre que sueña con la violencia, con el hierro y con el fuego, como único remedio para limpiar a su tierra de la miseria del pasado" (No, I am not a liberal; I am just a man of my time, exactly as my country's circumstances have shaped me, and not as books show. I am a Jacobin; a man dreaming with violence, with iron and fire as the only medicine to clean his land of the miseries of the past).[19] Allusions to iron, fire, and work synthesize in this passage the allegorical referents that articulate Blasco's social foundry, an assessment that unveils a submissive society in need of revolutionary action to transform the country.

Aresti is not the only character in the novel with a progressivist perspective. Foundry engineer Fernando Sanabre, a young man whose liberal ideas resonate with Blasco's political interests, is the novel's counterpoint to a young generation of conservative, Catholic, and nationalist professionals. For him, the ideal of pursuing the common good is above any religious mandate: "Sí, vida mía, tengo religión—dijo evasivamente. Creo que el hombre debe ser bueno y feliz sobre la tierra y para ello trabajo" (Yes, my love, I have religion—he said, evasively. I believe that men have to be good and happy on this Earth, and I work to contribute to this goal).[20] This vision contrasts with the capitalist principle that seems to rule Bilbao's economy in the novel, according to which particular interests prevail over collective needs—ambition and greed, as a consequence, overshadow social consciousness in favor of capital and power accumulation. Sánchez Morueta's submission to such doctrine represents in that regard the moral and intellectual failure of the industrial bourgeoisie as agent of social reform.

In fin-de-siglo Spain, industrialists were considered to encompass the highest human qualities. In fact, many handbooks of instruction for elementary school include descriptions that emphasize their work ethic, intelligence, honesty, and even their physical qualities.[21] All these merits, however, were completely blurred in the actual businessmen. Contrary to that idealized depiction, in most cases industrial tycoons allowed the oppression of the proletariat. If one adds to this situation the Church's interference in social dynamics with the purpose of mitigating workers' unconformity and favoring exploitation, it is easier to understand Blasco's pessimistic assessment. As he shows in the novel, miners' social unrest was in part instigated by an evident falsification of Christian ideologies conveniently embraced by the elites to justify inequality and suffering. As cultural critic Roland Forgues notes in his reading of the novel, this logic became ineffective as younger generations of workers started gaining a better understanding of their key, indispensable role within the industrial economy. Thus, while older workers accepted the exploitation of their labor force in the name of Christian morality, the younger ones rejected Christian morality in the name of the exploitation of their labor.[22] It comes as no surprise, then, that some characters in the novel saw in the past a lost, idyllic harmony that needed to be recovered. That is the case of Jesuit's disciple and counterpart of engineer Sanabria, conservative lawyer Fermín Urquiola, who blames the growing social unrest on progress: "Lo que demuestra que los antiguos tiempos eran los buenos y que, para tranquilidad de todos, hay que volver á la época en que no había progreso y los hombres vivían tranquilos" (What this proves is that the old times were the good ones, and that if we want the peace of the past we need to return to the time when progress was not a thing and men lived peacefully).[23] In the novel, the Church manipulates its parishioners against industry's social transformation in order to facilitate the return of the past that Urquiola feverously defends. Through the confessional, for example, priests shaped ideologies and influenced people's decisions; in public, from the pulpit, they condemned liberal ideas to prove the preeminence of religion over progress, while exalting poverty and suffering as the only ways to achieve salvation.[24]

It is indeed this interventionism of the Church, and its strong antagonism toward the proletariat, that leads to the violent encounter depicted at the end of the novel. Based on real events, the confrontation is a re-creation of the clash between a massive demonstration of workers and a pilgrimage of devotees of the Virgin Mary that took place in Bilbao on October 11, 1903. In his account of these events, historian Juan Pablo Fusi underscores the violence of the clash, with political leaders and priests among the dead and injured.[25] Among the causes that moved the demonstration were the long workday and precarious working conditions, scant remuneration, and, overall, the extremely limited possibilities of social improvement, notwithstanding industry's evident economic boom. Inequality was incontestable in the Bilbao of this period: while the industrial bourgeoisie and the Church lived in wealthy conditions, exhibiting luxury and comfort, workers endured extreme poverty and were deprived of appropriate living conditions. Years earlier, in an article entitled "El socialismo bilbaíno" (Bilbao's socialism), published in the literary journal *Germinal* on July 16, 1897, Ramiro de Maeztu already forewarned his readers of this situation: "De un lado, la minoría afortunada que levanta, para recreo de sus ocios, los hoteles coquetones . . . y las mansiones espléndidas del magnífico ensanche de Bilbao. Del otro, la mayoría de desventurados, guareciéndose en esa cuenca mineral cuya fealdad infunde espanto, y partiendo la vida entre el sombrío hormiguero de la mina y el barracón inmundo" (On one side, the fortunate minority who builds, for leisure, handsome manors . . . and splendid mansions in the magnificent Bilbao's modern district. On the other side, the majority of unfortunate people, taking shelter in a mineral basin whose ugliness inspires nothing but fear, and who divide their existence between the dark anthill that is the mine and their filthy shacks).[26] Later that same year, Maeztu would return to the topic in another article published in *El Heraldo* on October 28, entitled "La obra del odio" (The effect of hate). In the text, Maeztu blamed the workers' uprising on what he describes as industrialists' "scandalous exhibition of luxury."[27]

The reluctance to negotiate some of the proletariat's most basic demands—salary increase, reduction of the working day, adequate infrastructure for public services, and housing—could be in part explained by traditional society's fear of social change. Hence the persistence of feudal practices or their refashioning within the industrial economy. Moreover, since exploitation was understood as a necessary evil to sustain the dynamism of progress, workers' living conditions did not represent a major concern—their dehumanization was justified on the vision of industrial society as an enormous mechanism.[28] Allusions to industry as a mythical monster who fed on human labor follow a similar logic. Resorting to these images, Blasco's novel elaborates on what he perceived as the perversions of progress; it presents industrialization as a sort of new feudal economy that exposed the failure of the Restoration's political system to serve the actual needs of modernization.

Blasco's articulation of the industrial conflict, however, tends to be Manichean. The main characters in the novel are typified as supporters of either the working class's insurgent determination or the religious devotion and nationalistic tendencies of the Jesuits' acolytes. This strong polarization, nonetheless, is useful for exploring the violent encounter between tradition and progress, and for addressing the conflicts of regional nationalism. Set in the Basque country,

Blasco's novel incorporates ideas of ethnic, geographical, and religious identity. Given that the majority of the workforce in the local steel industry came from other regions, industrial growth was also considered a threat to cultural integrity. The Church capitalized on this population disparity to promote a specific stratification of society—one in which the proletariat was segregated for their assumed affiliation to liberal and progressive ideas, conditions that make them alien to an alleged *authentic* Basque identity. Underlining the political character of this conflict, historian M. K. Flynn has shown how social divisions also demarcated ideological lines: "The Basque oligarchy was loyal to the liberal monarchy; urban workers supported the far Left; and the Basque lower classes and rural population were to gravitate towards Basque nationalism."[29] In the novel, different characters show how these links among social, political, and nationalistic factors operated. Urquiola, for example, seems to defend modernization only to the extent that the society it promoted validated ethnic and regional distinctions:

> Que no le hablasen á él del populacho de las minas; corrompido y sin fe: hombres de todas las provincias, *maketos* llegados en invasión, trayendo con ellos lo peor de España, contaminando con sus vicios la pureza del país: siempre descontentos y amenazando con huelgas, deseando el exterminio de los ricos y comparando su miseria con el bienestar de los demás, como si hasta en el cielo no existiesen categorías y clases.[30]

> (He did not want to hear about the plebs of the mines; corrupted and irreligious; people from all provinces, foreigners coming to invade, bringing with them the worst of Spain, contaminating with their vices the purity of the region; always unhappy and threatening with strikes, hoping for the extermination of wealthy people and comparing their misery to the wellness of everyone else, as if even in Heaven there were no classes and categories.)

From this perspective, immigration was just another way in which industrialization had contributed to the contamination of an otherwise *pure* region—newcomers' non-normativity made them dirty, miserable, and evil, whereas locals remained clean and wealthy.

Yet, according to Blasco's social foundry, these nationalistic feelings originated in misinterpretations and misrepresentations of a so-called regional essence, and as such they were visible symptoms of the country's problematic modernization. In the novel, Aresti makes this point clear:

> Vizcaya no tiene apenas historia . . . y por esto posee la energía de los pueblos jóvenes. Su grandeza empieza ahora; sólo que los enemigos de lo moderno no lo ven. Su gloria es reciente y está en la ría, en el puerto, en las minas y las fábricas, en los buques que pasean por todos los mares la bandera de su matricula, en el esfuerzo colosal de dos generaciones que han trastornado la naturaleza para explotarla. . . . Este es un país que no ha dado en los tiempos

> pasados más que obispos y marinos. Ahora despuntan los únicos hombres notables que puede producir esta raza con sus especiales condiciones.[31]
>
> (Biscay barely has history . . . and because of that it possesses the energy of the young nations. Its greatness is beginning now; the problem is that the enemies of modernization are unable to see it. Bilbao's recent glory comes from its river, its harbors, its mines and factories, from the vessels that display our flag around the world, and from the effort of two generations who worked unceasingly on the transformation of nature to profit from it. . . . This is a region that in the past only produced sailors and priests. Now the only notable men this race has produced are finally excelling.)

Aresti points to the hypocrisy of society—as a source of wealth, material modernization was welcome in the region; once its transformations became a threat to the traditional structures that supported privilege, however, industrialization was censured and rejected on the grounds of religious and nationalist incompatibilities. Questioning associations between a "glorious past" and regional pride, the novel criticizes a nationalism that looked at the past instead of praising the achievements of the present.

Considering Catholicism to be an ideological tool that industrial elites used to justify and sustain their exploitation, the proletariat also distrusted national sentiments that were linked to religious values. Nationalists, in turn, saw workers as a threat to social stability and made them responsible for the fast deterioration of moral and religious traditions.[32] This form of nationalism took advantage of the strong animadversion between the Catholic middle classes and workers, and based many of its ideas on a contrast between the harmony of the past and the industrial dynamism of the present. The social foundry advanced by the novel is a reflection on the future of industry under such conditions. By exposing the negative consequences of misappropriating progress to sustain social structures incompatible with the idea of a modern nation, *El intruso* represents an evolution in the use of literary discourse as a socially committed cultural form. The texts analyzed in the next sections point precisely in that direction, placing the *labor question* at the core of modernization's social and cultural conflicts. Considering industry and workers as elements in a fragile and unstable reality, Manuel Ciges Aparacio's chronicles and Concha Espina's novel complement and problematize Blasco's reading of society, expanding on the particular dynamics that consolidated between the Church and the proletariat at the turn of the twentieth century.

Present-Day Society and the Limits of Fiction in Manuel Ciges Aparicio's *Las luchas de nuestros días*

In 1906, journalist and writer Manuel Ciges Aparicio traveled to the Mieres province in Asturias to cover the social unrest on the rise in this active industrial region. Reporting for the newspaper *España Nueva*, he followed a series of strikes and demonstrations organized by miners to protest corporate measures that were drastically affecting their living conditions. Ciges decided to go undercover to witness firsthand the unjust conditions that a small industrial elite had created to maintain their social status. He would inadvertently become an eyewitness to what later was known as the Mieres's big strike (*huelogona de* Mieres), an event that would expose alliances between the local upper classes and the Church with the purpose of discouraging, through the use of religion, workers' participation in the protests. In spite of the multiple formal complaints, the public demonstrations, and the violent strikes (one of which was the *huelgona*), conditions of exploitation in the region remained unchanged—workers were defeated by the greed of a minority. In fact, a couple of years later, in 1908, Ciges would have to cover again the workers' uprising, discontent that by then had spread to other industrial hubs such as Bilbao, Huelva, or Almadén. In his visit to these sites, the author saw the strategies of manipulation that banks and foreign companies implemented to keep in place their exploitative practices. Writing for the newspaper *El Mundo* (The globe), on this occasion Ciges completed a thorough report on the social phenomenon's various angles. This work, however, would never see the light due to the strong influence that the companies criticized in it had on the government of the time. Since these corporations were foreign multinationals, such an act of censorship not only represented an affront to social justice but also a challenge to national sovereignty.

This alarming, and in many senses scandalous, situation of corruption at the highest levels of power constitutes the center of these two chronicles, later compiled and finally published in two volumes with the suggestive subtitle "Las luchas de nuestros días" (These days' battles). In 1908, under the title *Los vencedores*, the first part of the reportage saw the light. The text received mixed reviews as it was mistaken for a fictional story, a situation that did not prevent Ciges from publishing, two years later, the second part of the work, *Los vencidos*, a text hailed as a milestone in what later came to be known as social journalism.[33] This section offers a close reading of some passages of both texts to highlight the narrative strategies Ciges uses to capture the conflictive nature of industrialization. As it elaborates a *clinical* diagnosis of the country's situation and looks for mechanisms to treat the illness afflicting it, this social foundry can be framed within the regenerationist narrative. In this effort, Ciges resorts to a nonfiction genre as a more valuable representation of modernization, thus challenging the prevalence of realism and the idealization of modernism, two of the most common ways literature approached reality during this period.

Commenting on Ciges Aparicio's archetypical role in the aesthetic turn of literary and political discourse toward a militant, more committed cultural expression, José Álvarez Junco has shown how this orientation would prevail for most of the twentieth century, turning with special interest to the mining industry as a paradigm of modernization's exploitative forms of labor.[34] It comes as no surprise that in his work Ciges recurrently references Blasco Ibáñez—their portraits of social disparity and the particular attention paid to the bourgeoisie's moral decline formed part of a larger project to evaluate Spain's conflictive social and economic transformation. Both writers thus concurred in their assessments: Cánovas's political system encouraged and even facilitated the distortion of modernity's welfare goals in favor of a conservative, outdated idea of Spain. Victor Fuentes, a renowned Ciges scholar, also underscores this connection with Blasco and shows how both authors aimed at exposing through their work the political corruption that prevailed between 1893 and 1923—the thirty years during which Cánovas's system operated.[35] In contrast to Blasco's literary works, which romanticize or dramatize industrial reality, Ciges's investigative pieces depict social actors and situations as they really were. Characters that in literature play a prototypical role, such as the businessman, the doctor, the engineer, the priest, the nationalist, or the worker, are in this case actual actors whose position in a defined historical moment highlights the hypocritical posture of certain sectors in society.[36] Furthermore, the chronicles' narrative structure tends to be linear and repetitive—they describe the particular context in which the situations or conflicts take place, then transcribe the interviews between Ciges and his interlocutors, and from there they analyze the negative impact industrialization had on people's lives. All these elements consolidate a social foundry through which the author re-creates, for audiences disconnected from the industrial world, some of the cultural negotiations and social struggles of capitalism.

One of the most detrimental effects of industry was the workers' dehumanization. In Spain this exercise of power and control operated in two distinctive ways, which Ciges documents with alarm and perplexity in his work. As noted earlier, under the new efficiency parameters set by the implementation of automatic and mechanical processes, understandings of time drastically changed. In industrial environments, the capitalist premise *time is gold* determined the configuration of social interactions and the rhythms of daily life. This new conceptualization was linked to notions of *utility*—workers were useful insofar as their output capacity or efficiency during a specific period of time was optimal, namely, as the cost of their work was less than the profit they could generate. Discussing this form of evaluating the worker's performance, Ciges comments: "Y como el tiempo es oro para un inglés, y la Compañía no gusta de perder el oro, le dirán por la noche que sólo ha realizado tantos viajes, que es inepto para el servicio, y poniéndole la cuenta en la mano, le enviarán á su casa" (And since for an Englishman time is gold, and the company would not like the idea of losing money, by the end of the workday they will tell the worker that he has only completed a certain amount of work, and that, consequently, he is not apt to offer his services to the company, and after paying him for his work, he will be fired and sent home).[37] Companies assumed workers to be just numbers in their accounting books, an attitude confirmed in the despotic character of practices that, paradoxically, were *out of phase* with the progressive spirit of modernization.

This dehumanization was also evident in the mechanization of labor. From an industrial perspective, workers were seen as sorts of automata, devices conditioned to respond to specific commands and operate within certain guidelines. Ciges witnessed with perplexity how miners followed orders communicated through the sounds of a siren: "Allá enfrente vuelve á sonar la bocina, y los trabajadores salen de sus escondites formando en un banco larga fila. La nube de pólvora y polvo se ha disipado, y en el lugar de los barrenos se ven bloques amontonados, paredes desgarradas" (There, in the front yard, a siren produces a sound again and workers leave their hideouts forming a long line in a bench. The cloud of gunpowder and dust has dissipated and instead of boreholes, piled blocks and torn up walls become visible).[38] The image of destruction that follows the gunpowder explosions mirrors in some way that of the workers, who, aligned with "torn up walls," exposed their decay and depersonalization. By symbolically associating labor oppression with mineral exploitation, in these descriptions industry's destructive effects transcend the landscape to impact the body.

Yet, notwithstanding the level of degradation that workers endured under these conditions, companies somehow managed to contain rebellion, making no changes to their operation in spite of the generalized and growing discontent. Baffled with this situation, Ciges scrutinized the reasons to discover that they lay in the memory of tragic events. In 1888, Rio Tinto, a multinational British mining company that controlled the iron exploitation in the Huelva region, had contained the workers' insurgence through the use of military force. This brutal response involved the state army and left hundreds of workers dead and many more injured. One of the miners interviewed by Ciges recounts the event in these terms: "4 febrero 1888. Esta es la fecha terrible que se repite cien veces diarias y que pesa como una obsesión Hasta los que no asistieron á ella la recuerdan con invencible temor, pues en fuerza de oírla repetir, la imaginación se la representa con todo su trágico vigor de traidora hecatombe" (February 4, 1888. That is the terrible date that every day we obsessively repeat in our minds one hundred times. Even the workers that did not attend that demonstration recall it with invincible fear, for listening to the story once and again triggers their imagination, exposing the event's tragic condition of treacherous determination).[39] The way in which workers' demands were negotiated in the past changed their expectations in the present; Ciges continues: "Hoy sólo creen en la eficacia de la violencia. Aislados viven; á sus trabajos van, y en sus casas se meten luego" (Nowadays workers only believe in the efficacy of violence. They live in isolation; they go to work and return home to stay there).[40] Accumulation of hatred and resentment had become part of the fuel that fed industrial dynamism. If violence was believed to be the only mechanism for channeling that force, then it comes as no surprise that most of the industrial enclaves were under siege, waiting with fear and distrust for this volatile situation to explode. Ciges's social foundry points to an inflexible class stratification and other feudal-like practices as the main causes for this situation.

Looking at heat and energy as ancestral forces that emphasized the destructive capacity of industrial practices, Ciges's description of the labor space is full of images of greed, death, and torment: "Al fin del monte, junto á la vía férrea, se yergue la fundición de Bessemer, y aunque está bien guardada, basta pasar al lado para presentir su medroso interior. Es como las fraguas de Vulcano ó como un rincón en el reino de Plutón" (At the bottom of the hill, on the side of the railroad, there is the Bessemer foundry, and even though it is very well guarded, it is enough with passing by to imagine its frightening interior. It is like Vulcan's forge or like a hideout in Pluto's kingdom).[41] Mythological references here have a purpose that is not necessarily spiritual; rather, they are aimed at highlighting specific material conditions that associated industry with ancestral images of the underworld.

In his two chronicles, Ciges elaborates a painstaking description of the social problems around the mining industry in the region; yet in his account the role of the state is merely nominal. This absence underlines political, economic, and nationalistic tensions that resonate with the mining companies' practices and operational models. Bestowed with the autonomy to expropriate natural resources, multinational corporations tended to dehumanize labor to justify their ruthless exploitation of both workers and nature. Disregarding the paramount importance of the proletariat in its projects of national modernization, the state turned a blind eye to their problems and demands. From Ciges's perspective, lacking a system to protect workers from abusive industrial practices was unacceptable. Feelings of shame and anger at the inhumanity that prevailed in the mining complexes, the companies that produced it, and the state structure that allowed it thus shape his denunciations. The descriptions of the miners' living quarters, for example, are a testimony of these emotions: "Entro en el cerco de Buitrones, el cerco siniestro donde está la fundición, y, al llegar á una puerta, quiero retroceder avergonzado. . . . ¡Yo siento profundo amor á España, y en este momento me avergüenzo de ser español!" (I cross the fence of Buitrones, the sinister enclosure where the foundry is located, and as I approach the door, I want to pull back in shame. . . . I feel a profound love for Spain, but at this moment I am ashamed of being Spanish!)[42] Within this context, the workers' response of resistance, confrontation, and conflict seems justifiable, exposing, in addition, the irreconcilable character of two different approaches to industrial modernization. On the one hand, there was the perspective of foreign companies and investors, who tried to force outdated economic and social practices on the industrial dynamism; on the other, the viewpoint of workers and liberal sectors, who saw in industrial modernization an opportunity to positively transform the country.

Violence, however, was more than just a manifestation of the workers' unconformity; it was also a metaphorical exteriorization of the consequences of corporate imperialism—workers had been *colonized* by industry while the country's wealth was expropriated by foreign conglomerates. From the companies' perspective, resistance, both political and ideological, was categorized as illegitimate and subversive. In the workers' view the situation was different—not only did they have the right to protest, but their condition demanded a vigorous response: "En España sólo se puede ser anarquista. Todo propende á hacer anarquistas. A los socialistas se nos tiene por un partido demasiado afecto a la legalidad, y así es; pero los gobiernos maldito si han sabido agradecérnoslo" (In Spain one can only be an anarchist. All conditions are set for workers to become anarchists. Socialists are considered a political party particularly predisposed to legality; and that is true, but the damned government has never recognized this predisposition).[43] The legal framework under which different actors in this conflict were held accountable was inconsistent with industrial reality; the state's weak presence in the region was to blame for the gaps that allowed or even prescribed the use of violence in containing the workers' insurgence.

Ciges's negative assessment of industrialization opens the door to reminiscing about and idealizing the past. In the chronicles, for example, rural spaces are portrayed as peaceful and harmonic. Demonizing science, technology, and industry with the purpose of ennobling the past, descriptions in the text resort to tropes usually deployed in works of the period to textually negotiate the impact of modernization. Following this logic, a contrast between past and present is created in the chronicles to mark industry as a turning point in people's lives: "Nadie se quejaba; la agricultura, que proveía antes a las necesidades del pueblo, se arruinaba ahora; los ríos se enturbiaban con los arrastres del carbón y ahuyentaban la pesca, que había sido la segunda fuente de riqueza; los hundimientos de las galerías deprimían la superficie de los montes e inutilizaban los campos" ([By then] no one complained; agriculture, which supplied all of the town's needs, now was getting completely ruined; the rivers were spoiled with coal residues that prevented fishing, an activity that had been the second most important source of wealth; sinking underground passages created depressions on the surface of valleys and mountains, making lands unusable).[44] Yet this immense material and symbolic loss was neglected by the government and disregarded by those who had the resources to alleviate it—the system kept running, and with time it only produced more erosion and injustice.

Nature's deterioration not only mirrored that of the workers, but it also exposed the moral degradation of social institutions. That was the case of the Catholic Church and the role it played in the tension between exploited and exploiters. Many of the miners that Ciges interviewed agreed on that point: "Su único oficio [el de la Iglesia] es hablar, intrigar, acechar los hogares donde puedan rendir almas y obtener beneficios" (Its [the Church's] only function is talking, intriguing, stalking those homes where they can submit souls and obtain rewards).[45] Here, Ciges's criticism of the Church's double standards is similar to the one Blasco presents in his novel. In *El intruso*, religion, represented by the Jesuits, infiltrated the private sphere of the higher social classes to manipulate their view of industrial modernization, thus profiting from the wealth generated at the workers' expense. Ciges, in turn, offers evidence of similar strategies employed within the miners' hearth, where the Church invaded privacy to destabilize the family structure. This interference in the workers' universe also exposed a disquieting reality. Traditionally, the Church had opposed liberalism, since it was perceived as a threat to power structures. Amid industrial expansion, workers became part of this menace to the stability of the past. Yet the institution expressed disapproval or endorsed the demands of the working class depending on its particular interests. This ambiguous attitude translated into people's distrust of religion and rejection of Catholicism. Ciges agreed with many of his interviewees that the Church's main interest in the labor question was part of a strategy to manipulate and control the population in order to preserve preindustrial mechanisms of wealth accumulation and social stratification.

But even though the Church's manipulative agenda was evident, in most cases devotion and respect to religious authority prevailed over determination. Priests' constant intervention in the private realm was not only difficult to avoid, but it had been somehow normalized. Companies were aware of their employees' religiousness and used it to contain insurrection from within; as a miner explains to Ciges, "Es tan burdo este juego entre la Religión y la Fábrica, que todos los hombres lo vemos; pero no lo ven las mujeres, y eso es bastante para que triunfen" (This second-rate ploy between religion and the company is so manifest, that we all, the workers, see it; but not our wives, and that is enough for them [the Church] to succeed).[46] Companies also used workers' devotion as a control mechanism—knowing that religious workers were less inclined to promote or participate in insubordination, managers and supervisors attended services to pay particular attention to the parishioners. Processions and ceremonies thus became public spaces of surveillance through which it was possible to identify those workers who represented a potential risk. Flagged employees were usually fired without justification: "Casi todos los que derrotados y expulsados tras la huelga quedaron en la miseria . . . contemplaron con tristes ojos de hambrientos el desfile de la gigantesca procesión. Los que en ésta formaron serían esclavos, como usted dice; pero tenían el alivio de poder comer" (Almost all the defeated and fired miners who were pushed to live on the breadline after the strike . . . contemplated the giant procession with the sad eyes of hunger. Those who formed part of it were going to continue being slaves, as you have said; but at least they were relieved by having something to eat).[47] This mechanism of segregation became an important variable in Ciges's evaluation of the contradictory nature of progress. For him, the various strategies of social control employed by companies were just a symptom of a larger class struggle—one led by the industrial bourgeoisie—to negotiate multiples facets of progress and its transformations.

As in the case of other authors studied in this book, Ciges's chronicles present an accurate diagnosis of a dysfunctional society that seemed unable to synchronize its social and economic aspirations with modernization's particular dynamism. One of the visible results of this conflict was the emergence of exploitative labor practices that tried to impose a feudal logic on the industrial economy. These two texts censure the mechanisms of labor exploitation that large corporations put in place with the complicity of the Church and the negligence of the state. As testimonial evidence of the complex assimilation of new realities, the two pieces expose the different mechanisms of manipulation that the elites implemented in order to contain social change. The Church's demonization of socialist ideas and stigmatization of the working class as threats to traditional orders are just two examples of such (mis)appropriations of industrial progress—the testimonies compiled in *Los vencedores* and *Los vencidos* thus show the difficult ideological balancing of tradition and modernity. Ciges concludes: "El ansia del oro abandonó a la madre tierra para explorar las entrañas de los montes, más negras que las entrañas de los condenados. Los campos se olvidaron y la mala hierba los pobló" (The desire for gold left behind mother earth to exploit the mountains' entrails, a space darker than the entrails of the damned souls. [In the meantime,] the lands were forgotten, and weeds took over them).[48] Ciges's social foundry highlights the proletariat's defeat in vindicating its rights, instead presenting modernization as fertile land for the defeaters (the industrial bourgeoise, the Church) to take over—like the weeds described in the passage.

In all this process, the Church's ambiguous role stands out: in Ciges's view, its alleged moral and social commitment to workers in reality hid an agenda to favor their subjugation. The next section analyzes a different take on this problematic relationship. For writer Concha Espina, the Church's engagement with questions of social inequality and labor exploitation was authentic and indispensable. Catholicism offered a space of conciliation between workers and corporations wherein modernization and faith, past and present, tradition and progress could converge and even harmonize. Her social foundry thus reveals yet another perspective on the complicated assimilation of industrial progress into the country's collective imagination.

The Social Cause of Catholicism in Concha Espina's *El metal de los muertos*

With the promulgation in 1891 of the papal encyclical *Rerum Novarum*, the Catholic Church acknowledged the urgency of assuming an active role in the social challenges posed by industrial modernization. Even though this concession did not imply an open acceptance of socialist principles, the creation of missions to serve impoverished communities in industrial areas demonstrated Pope Leo XIII's interest in addressing social change, as well as the Church's approach to the problems of the growing working class. In Spain, conservatives received this gesture with some skepticism and suspicion: was the Church, an unconditional ally of the elites in the administration of society, betraying their trust? A closer look at the encyclical, however, shows that was not the case. In fact, the document's exhortation to create task groups to work on alleviating tensions between capital and labor did not respond to progressivist principles; on the contrary, it was a conservative and doctrinal initiative conceived with the purpose of preserving traditional structures of wealth accumulation and social segregation without neglecting the workers' need of basic levels of welfare. The goal of the pope's text was not that of proposing an ideological revolution within the Church, as some sectors of society understood it at first, but offering a set of guidelines on how the institution could get involved in the different social conflicts of modernization.

Many intellectuals of the time endorsed the encyclical's paternalist approach. A devoted Catholic, Concha Espina was an active part of this group, as it is evident in her novel *El metal de los muertos* (1920). Her approach to the problems of the proletariat, accordingly, differs from that of Blasco or Ciges. These two writers, as noted earlier, saw the Church's interest in social problems with suspicion—for them, religion was a tool of ideological manipulation that helped sustain a nefarious system of exploitation. The Church, in this vision, was in part responsible for the precarious situation of the proletariat. For Espina, instead, Catholicism and religious devotion offered mollifying spaces in which the workers' suffering was justified as a necessary but rewarding sacrifice. In her novel, injustice is not presented as the direct result of a system of social stratification, whose structure and limits the Church eagerly defended, but rather as the result of international companies meddling in the Spanish economy. This interference was not only a violation of Spanish economic and political sovereignty, but also an obstacle to the renegotiation of identity in the context of industry. The tension between workers and corporations was, in the end, a struggle to defend national traits, and as such an ideological battle on political, cultural, and social fronts. Espina, in any case, was aware of the Church's decision to side with the oppressors amid the labor conflict, and she reproved the industrialists' particular embodiment of Catholicism. Consequently, the revolutionary ideal defended in the novel remained Christian in essence but did not necessarily subscribe to the Church's institutional stance on the social question.[49]

If Catholicism was a distinctive feature of the Spanish identity, its perversion, namely the distortion of its social justice principles on behalf of alien economic interests, was also the corruption of the *national* essence. The novel's structure and argument support this vision. *El metal de los muertos* is the story of two couples whose lives intersect in one of the many workers' revolts that took place in Huelva at the beginning of the twentieth century. The first couple, Cantabrian villagers Gabriel and his fiancée, Aurora, abandon their lives as fishermen to migrate to the south in search of better economic opportunities. The second duo, committed social activists and siblings José Luís and Rosario Garcillán, has relocated to Huelva in a Catholic mission to offer aid and moral support to victims of the mining industry's unjust systems of labor exploitation. The encounter between these two couples allows Espina to elaborate a complex portrayal of the growing social tensions in industrial Spain. Her social foundry is a condemnation of the country's dependency on and submission to foreign models of economic modernization, and of the roles that the Church and the state played within this process.

In the novel, Espina employs complex metaphors and allegories based on industrial imagery to call attention to how companies profit from natural as well as human resources—the essential origins of national wealth. Mining in that context functions as a backdrop against which past and present can be contrasted. Gabriel and Aurora's relationship with the ocean in the past and their present subjugation to the dynamics of mineral extraction, for example, serve as powerful referents to describe reality. Given the impoverishment of rural areas, for the young couple *descending* into the subsoil is much better than making a living on the surface: "Cobró aliento el muchacho al abrigo generoso del amor; logró que le recomendasen al ingeniero de una mina cercana, y una tarde huyó de la costa buscando *en el fondo de la tierra* el auxilio que le negaba el mar" (The young man gained energy from true love's generosity; he managed to get a recommendation to present to an engineer in a nearby mine, and one afternoon left the coast to seek *underground* the sustenance that the ocean had denied him).[50] In his journey from the ocean to the land, from the surface to the subterranean mine, Gabriel will have to confront industry's insatiable hunger—instead of feeding workers, modernization consumes them. Here the image of *industry as a man-eating monster* is complemented with sensorial metaphors; machinery's loud sounds, for example, turn into a new kind of music whose harmonic construction reflects the destructive character of progress:

> Es un enorme lavadero de mineral, un vertiginoso mecanismo que rueda y gira, choca y ruge con incesante clamor. Las mandíbulas de hierro devoran la calamina y la escupen en los coladores monstruosos para echarla después a un nuevo tamiz, a un inquieto canal, a un trémulo cajón. Aturden el vaivén continuo de las cribas, la rotación constante del Tremel, que igualan el material, y la trepidación de las mesas giratorias, el crujido de las ballestas y los ejes, el movimiento de los tambores y las planchas: zunchos y rodillos, cables y zubias, el aire y el polvo, el agua y las piedras tienen ritmos, gestos, curvas, fauces, una delirante voz de orgullo y de poder, la tremenda armonía de un verso torrencial.[51]

> (It is a huge mineral-washing device, a vertiginous mechanism that rolls, turns, and roars with an endless groan. The iron jaws devour the calamine and then spit it over monstrous strainers to later place it in another filter, in a restless canal, and then in a trembling crate. The swinging of the sieves, the constant rotation of the Tremel mill to mix the materials, and the trepidation of the turning table, the creaking of the crossbows and the axes, the movement of the cylinders and the plate are also stunning: metal bands and rolling pins, cables and gears, air and dust, water and stones have particular rhythms, gestures, curves, openings; a delirious voice of pride and power with the tremendous harmony of a torrential verse.)

Machines embody a power that is above men, one that "deliriously" challenges the superiority of God. Industry, in foreign hands, antagonizes not only social justice by exhausting human capital, but also religion, by imposing the materialism of progress over the spiritual essence of national identity.

Espina's abstraction of society as a mechanical system and her depiction of material reality in human terms signal the reversal of ethical values associated with Catholicism. While workers were assumed to be mechanical pieces in an enormous engine, machinery and nature are anthropomorphized to highlight the spiritual components of modernization. In line with a nonessentialist representation of material progress, Espina reminds readers that industry's capacity for creating is bigger than its destructive power: "Bajo los muros dislocados y la cimera tortuosa adquieren las piedras formas humanas, igual que en el taller de un escultor" (Under the misplaced walls and the tortuous crest, stones acquire human shapes, as happens in the sculptor's workshop).[52] Being a mythical force, industry is able to (re)shape nature and society with artistic ability. Described as spiritual quests in the novel, strikes to protest capitalist excesses are not aimed at rejecting a particular social order but at vindicating national values: "No se trata sólo del incumplido régimen del trabajo ni de la perpetua lucha entre jornaleros y patronos: hay sobre ésta una previa cuestión nacional" (It is not just a matter of unfulfilled labor agreements or a response to the tension between employers and workers: beyond all that there is mainly a nationalist issue).[53] From Espina's perspective, the state's relinquishment of its natural resources was illogical, for the wealth they produced was enough to improve the entire social system. By allowing foreign companies to ignore the principles of prosperity on which progress had been predicated, the government seemed to be recanting the foundations of Christianity to favor individual greed over collective justice.

Espina's social foundry confirms these distortions by rendering nature as a deformed, almost monstrous space. If modernization had deviated from its goal of creating the conditions to guarantee society's well-being, progress could then be seen just as an expression of individual success, a celebration of corporate greed: "Centenares de años en activa industria consiguieron desollar cimas y laderas, hundir quebradas, ejercer en valles y lomas inmensas depresiones, cegar de cadáveres los minados, entorpecer con espantosa negrura toda la vegetación" (Hundreds of years of industrial activity led to the complete defoliation of peaks and slopes, buried ravines, created enormous depressions on valleys and hills, filled mines with corpses, hindered vegetation with horrible blackness).[54] The impact of mining on the environment had already caused a profound impression on early explorers of the peninsula, such as Rodrigo de Caro (1573–1647), whose astonishment is highlighted in Espina's own reflection: "¿A quién no admira ver que el atrevimiento humano osase a tanto y que fuera más dura el hambre del oro que la dureza de las peñas?" (Who is not amazed to see how daring human audacity is, and how the hunger for gold could be stronger than the hard mountains?)[55] This same rapacity, that in part justified the dehumanization of labor, transcended the limits of the mine and the factory to affect all participants in the industrial economy.

The violence and marginality typically associated with miners are transposed in the novel to prostitutes, highlighting yet another negative, transgressive aspect of modernization: "Allí viven con la más escandalosa brutalidad; duermen al raso, comen relieves de los alimentos mezquinos trasegados en las serillas de los caminantes, reciben injurias y golpes a trueque de unos céntimos y unos mendrugos, y no es raro que alguna quede muerta de una paliza o de una puñalada en los atroces holgorios" (There, they [prostitutes] live in the most scandalous brutality, they sleep in the open, eat leftover food that travelers carry in their baskets, receive insults and abuse in exchange for dimes and crumbs, and it is not unusual that some died from being beaten or stabbed in these atrocious parties).[56] Espina's social foundry operates as a moral canvas wherein social actors are characterized in Manichean terms. The Garcillán siblings are a good example of this oversimplification and a counterweight to the pessimistic depiction of prostitution—members of an idealized, generous and kind-hearted bourgeoisie, they are the perfect prototype of a social class that in the author's view embodied the principles of justice and harmony of a modern Christian society:

> Rosario y José Luís conocían la tristeza de muchos barrios obreros, alfoces insalubres de las grandes industrias en el mundo, puestos como un halo de martirio y de sombra en torno a la soberbia y al poder; mas no concebían en España una multitud infeliz, un trágico reino de los pobres que ofreciera el contraste de su dolor para marco de la injusticia humana, precisamente bajo el cielo más altivo y azul del país, y sometida a patronos extranjeros.[57]
>
> (Rosario and José Luís were familiar with the sadness of many working-class neighborhoods around the world, industry's insalubrious slums, auras of martyrdom and darkness that highlight arrogance and power; yet they could not conceive the idea of an unhappy multitude in Spain, a tragic kingdom of

poor people that offered its pain as a backdrop to contrast human injustice against the brightest and bluest sky in the entire country.)

The idea of martyrdom as a sacrifice that served capitalist interests highlights industry's distortion of morality and religion. For Espina this was an inconceivable reality in a country like Spain, where nature had been generous enough to produce immense wealth.

In Espina's assessment, the bourgeoisie emerges as a force able to recompose progress's ethical values. Besides the Garcillán siblings, other characters in the novel assume this task. If modernization was understood as a sort of moral disease that affected national identity, then medical doctors would be ideal carriers of the author's message of spiritual healing. Indeed, as a practitioner of medicine and a man of science, the doctor in the novel holds a privileged objective position from which to approach reality: "Es Alejandro Romero, el médico de los trabajadores, un sevillano alegre y bondadoso, que al terminar la carrera fue a prestar sus servicios al Sindicato nervense, inclinándose a la causa de los obreros bajo una porción de aficiones morales y desinteresadas. Era romántico y poeta, con un espíritu muy religioso, atento a las honduras de la vida, propicio a los ideales extraordinarios" (This is Alejandro Romero, the workers' physician, an optimistic and kindhearted man from Seville, who, after completing his studies, went to serve the Union of the region, leaning toward the workers' cause without any particular interest except his own moral perspective. He was romantic and a sort of a poet, with a very religious spirit, attentive to the complexity of life, and fond of extraordinary ideals).[58] Poetic sensitivity is presented here as an essential asset in a doctor who needed to interpret the symbolic diseases that modernization had inflicted on society.

Another strategy used in Espina's novel to denounce industrialization's social disarrays consists in projecting this negative impact on the landscape—the *physical body* of the social organism. Through this metaphoric exercise, the text shows the extent of the damage that greed, foreign intervention, and excessive materialism had caused:

> Por cada absceso hinchado y caliente influyen las herramientas, lendeles y maquinarias a modo de bisturí; los estratos son como gruesas capas conjuntivas que dejan entre sí hendiduras donde las células se corrompen. De las zonas de mortificación surten médulas rosadas, azules, amarillas, con todos los matices cambiantes del cobre florentino. La tierra, como una piel orgánica y flexible, se levanta en senos y tumores, se cubre de manchas lívidas, de lunares morenos, de torvas cicatrices; se rompe en flujos sanguíneos, en heces gangrenosas, en licores serosos. Las corrientes vitriólicas sirven de melena a la terrible enfermedad; los escombros y residuos son la escara, la porción caduca del cuerpo que agoniza; los derrumbos se pueden confundir con ablaciones, y están rojas las cavernas de los vasos lo mismo que en las visceras

> humanas, opacas las túnicas de los alveolos igual que en el armazón del carcinoma. En la montera de los escalones hay quistes que exudan una papilla de color de herrumbre; las arrugias y los avenamientos rebosan líquidos hediondos como el pus. Los pliegues de la siniestra herida bullen comidos de gusanos: son los mineros que cavan su propia existencia, dejándose también devorar. . . . En torno a la llaga sinuosa ruedan los clamores del mineraje, y se levanta sobre ellos una voz hondísima, inconfundible, hecha de atropellos y hurtos, de caries y escisiones: ¡es la brama de los abismos, la pobre tierra moribunda que se parte en sollozos![59]
>
> (Tools and machines, like surgical knives, left a deep mark on every swollen and hot abscess; strata resemble thick layers of connecting tissue with fissures wherein the cells get corrupted. Pink, blue, yellow marrow spurts from the areas of hard labor, creating all possible shades of the malleable Florentine copper. Earth, as an organic and flexible skin, grows protuberances and tumors, gets covered with pale patches, dark moles, baleful scars; it breaks in bloody fluxes, gangrenous feces, serous liquors. Vitriolic currents work as the terrible disease's excrement; debris and waste are the crust, the expired portion of a dying body; landslides may be confused with ablations, and vessels' caverns are red, as if they were in human entrails; membranes of the alveoli are matted, like a carcinoma's shell. At the top of the stairs there are cysts exuding a rust-colored mush; tunnels and drainages overflow with fetid liquids similar to pus. The folds of this sinister wound boil while they are eaten by worms: these are the miners digging their own tombs, allowing themselves to also be devoured. . . . Around the tortuous ulcer, the crying of the mining industry can be heard while a deep voice raises, unmistakable, one made out of abuses and expropriation, of caries and excisions: it is the abyss's bellow, the poor earth breaking apart in sobs.)

Industry affected both nature and people, physical spaces and spiritual realms; symbolically, it could be seen as a disease that attacked the principles on which national identity was articulated. Espina's diagnosis points to the presence of foreign interests and the deterioration of Catholic values as the origins of this social cancer. A healthy nation, instead, would present an adequate balance between traditional values and material advantages and presuppose the country's repossession of its natural resources.

In the suffering society presented in *El metal de los muertos*, the proletariat's unconformity and their eventual uprising are only the symptoms of a troublesome assimilation of modernization. The invasive nature of foreign interests only complicated more this uneven, contradictory, and problematic negotiation. In Espina's perspective, the Church's active participation in this tension could be an opportunity, a means to somehow reconcile past and present. For they contested industry's capital sins, Catholic principles remained solid pillars on which the reconstruction of national identity seemed possible. A geological allegory illustrates this idea in the novel: "Vienen estas montañas del interior caliente de la tierra y se destrozan temblando, violadas por la codicia universal, abiertas al sol entre lamentos, como víctimas de un mundo. Sus torcales, sus andenes, cargados de pirita marcial, bajan a la llanura convertidos en un espeso manto rojo, una violenta pesadumbre que enloquece la vida" (These mountains come from the warm interior of the earth and, violated by universal greed, open to the sun amid laments, as victims exposed to the world, they get destroyed with a shock. Their cliffs, their paths, filled with martial pyrite, become roads covered with a thick red cloak, a violent sorrow that maddens life).[60] Because of greed, natural wealth produced suffering and violence, impoverishing the country while making other nations richer.

As cultural critic Elizabeth Rojas-Auda shows in her analysis of the novel, the extreme violence of the revolt depicted in the conclusion of the story only confirms that in the industrial world despair prevailed over justice and love.[61] The early optimism about industrial modernization that some authors convey in their work seemed to have faded by the time Espina wrote this novel. For her, there was no possible harmonization between religious values and modernity given the way in which industrialization had distorted the ideals and ethical principles of progress. Even though uses of science and technology to promote economic, political, and social change had in many ways challenged the basis and structure of national identity from the beginning, no author anticipated that the gap between tradition and progress was indeed insurmountable without the sacrifice of the working class. Violence was one of the expressions of this impracticality. However, and in contrast to Blasco's novel, in which traditional institutions are irredeemable, *El metal de los muertos* attempts a vindication of Catholicism as a positive actor within this conflict. Inequality and social confrontation, in any case, remained associated with the consolidation of the working class as a force, a potential that in the mind of many intellectuals, politicians, and educators could be capitalized and regulated through the use of scientific knowledge. In that context, public dissemination of science, as it is discussed in chapter 5, would become essential for the collective efforts to understand and assimilate the multiple transformations of progress.

CHAPTER 5

Technologies of Mass Diffusion

The publication of Charles Darwin's celebrated study on the origin of species in Spanish (1877) generated strong reactions among the intellectual community on the peninsula.[1] In a series of articles published the same year in the journal *La Ciencia Cristiana* (The Christian science), and later compiled under the title "Reflexiones científicas contra el Darwinismo" (Scientific reflections against Darwinism), for example, Spanish writer Emilia Pardo Bazán brands as absurd the ideas of the British biologist for they challenged the Catholic notion of human superiority over other species. As critic Thomas Glick explains, Pardo Bazán's articles emphasize the fact that Darwin's theory had been unquestionably accepted in Spain, a situation that was worrisome for certain conservative sectors of society.[2] Pardo Bazán's advice to her readers was to remain calm: although Darwin's ideas were hypotheses for a new scientific philosophy, they were not proved, uncontested laws. As a matter of fact, the author continued, *The Origin of Species* had not been as widely accepted as people in Spain tended to believe. On the contrary, important scientists and academics—French anthropologist Jean Louis Armand de Quatrefages or physician Jean Pierre Flourens, among others—radically opposed evolutionary ideas. According to Pardo Bazán, Darwin's popularity in Spain was owing to the wide dissemination of his postulations, which was largely attributable to the work of German scientist Ernst Haeckel. That these ideas were popular, however, did not mean they were true. Remarkably, her reflections, besides being an attempt to defend Catholicism while bringing one of the most important scientific debates of the period into the public debate, drew attention to the relevance of scientific divulgation in the country.

Two important trends show the role that journalism had in the assimilation of technological innovation, scientific discoveries, and industrial transformation in *fin-de-siglo* Spain. On the one hand, science communication decoded advanced concepts and ideas for the layperson, emphasizing their importance for society and their applications to daily life. It also contributed to creating allegorical devices to address social tensions at different levels. Through scientific and industrial images and abstractions, scientists, state administrators, educators, and the general public had a common language for talking about the country and its problems. On the other hand, the popularization of technological advances had a profound impact on people's imagination. Train travel, for example, changed collective perceptions about the country's geographical extension. In fact, the unheard experience of crossing the country in a short period of time—symbolically embracing the national territory—became one of the most important sources of literary inspiration for writers and journalists alike. This chapter analyzes examples from these two trends, showing first how Emilia Pardo Bazán and José Echegaray appropriated the language and images of thermodynamics, electricity, and the railroad industry to produce a particular evaluation of society. The second part discusses Pedro Antonio de Alarcón's travel account *Viajes por España* (Trips around Spain). I demonstrate how, by problematizing the incompatibilities of tradition and progress in the context of modernization, this report contributed to the assessment of national identity.

Even though nineteenth-century journalism was essentially a mechanism for the exchange of ideas, it also served as a vehicle for instructing readers. Many writers of the period considered the press to be society's *nervous system* and praised its ability to reach extended audiences.[3] Information and knowledge were resources that needed to be exploited, and the printing press was a matchless instrument for potentializing them. Rather than using industry as a metaphorical, allegorical, or conceptual referent, the *social foundries* discussed here are the result of the authors' direct experience with science or with its practical applications. Since it was precisely through the development of scientific knowledge and the mastering of technology that society had advanced, this close contact with the materiality of progress was advantageous to their assessment of reality. For these writers, it was clear that intellectual capacity, more than the potential for mechanical and manual labor, was key to social improvement. Inasmuch as higher education and technical instruction fulfilled their mission, coaching audiences through the popularization of science constituted yet another important component in the formation of the social body. It was in this spirit that chemist José Rodríguez Carracido commended Echegaray's work as educator and science communicator:

> La educación científica cada vez pone más al descubierto que todas las producciones de la actividad humana, aun las que no se logran sin extraordinaria destreza manual, deben ser antes preparadas por la cabeza que por las manos, asemejándose este proceso al del trabajo fisiológico, en el cual á la contracción del músculo ha de preceder el estímulo nervioso. . . . Según este

> orden, el engrandecimiento de un pueblo requiere en primer término el desarrollo de su vida intelectual.[4]
>
> (Scientific education every time reveals more about how all productions of human activity, even those that are impossible without manual dexterity, should be prepared with the head instead of the hands, in a similar way as in every physiological pursuit, muscular activity precedes the nervous stimulus. . . . According to this organization, the enhancement of a nation first requires the strengthening of its intellectual life.)

All the authors studied in the next sections highlight the importance of science and technology both as forms of knowledge and as embodiments of experience—for them, knowing and understanding recent discoveries, and the theoretical principles on which they were based, was as important as being in contact with them. In both fronts (theoretical and practical), spreading specialized knowledge helped audiences deal with the country's reorganization.

José Echegaray: Modern Theories of Physics for Out-of-Date Audiences

Echegaray stood out in many different fields; previous chapters have discussed some of his work on education, politics, science, and literature. As a popularizer of science, his aim was to make the most recent discoveries in physics accessible to a wider audience. This included topics such as astronomy, electromagnetism, and, of course, thermodynamics. In *Teorías modernas de la física*, a two-volume collection of essays published in 1873 and 1883, respectively, he not only describes and explains many of the basic premises of the theoretical apparatus supporting these discoveries but also offers insights into the importance of science, technology, and education for the country's development. Complemented by this work, his social foundry turns science into a common language to communicate many of his concerns about the social transformations of modernization and his expectations for the future. Echegaray's alignment with Krausist ideology was essential in this regard to highlight the importance of science without antagonizing his potential readership. His explanations of scientific empiricism, for example, are founded on the harmonious balance between reason and faith, and theory and practice. This particular approach sets the tone for his assessment of society and emphasizes the didactic nature of the entire text.

In a talk addressing the Royal Academy of Sciences during the induction of industrial engineer Francisco Rojas in 1894, Echegaray sets out the conceptual guidelines on which many of his explanations of physics for nonexpert audiences were supported. Departing from the notion of the unity of forces, he outlines the principles of universality and inviolability of physical laws, two fundamentals that operate on all scales in the universe: "Los fenómenos físicos obedecen á leyes comunes, lo mismo cuando se producen en gran escala y entre grandes masas, que cuando se realizan entre las más pequeñas á que nuestros sentidos pueden alcanzar" (Physical phenomena obey common laws, they manifest on a large scale and among large masses just as they affect the smallest things our senses can perceive).[5] If people had command of these propositions, they would more easily understand more complex phenomena. Convinced of the need of dissecting questions to find their elemental and better-known components, Echegaray grouped natural principles according to their relation to the foundational blocks of physics: matter, movement, and force.[6] Heat, electricity, magnetism, and light, the most important observable phenomena in the universe, could be understood through the study of these three manifestations and of the property that unifies them—energy. Nineteenth-century scientists had indeed focused on the study of energy, and more specifically on its materialization in the form of heat. In that context, as noted earlier, thermodynamics became the center of attention for both science and engineering, entering the collective imagination as the main conceptual engine of industrial development.

But if the unity of forces was an attempt to explain how the universe worked, its conceptualization had to be consistent with religion and, as such, include a spiritual component. As a devout Catholic, Echegaray did not understand science as a merely materialistic tool for explaining reality. For him, science and religion addressed different, but complementary questions about the universe. Aware of his audience's perception of science as a potential enemy of faith, Echegaray comforts the reader from the beginning of his text by acknowledging the existence of fundamental questions to which science did not offer solutions: "Podrá la física explicarlo todo con los átomos y el movimiento, pero no explicará ni el movimiento ni los átomos; ni de dónde viene, ni cómo es eterno ese oleaje infinito de lo material" (Physics might be able to explain everything through studying atoms and movement, but it won't be able to explain the atoms or movement itself; nor explain where matter comes from or how it constitutes the eternal and infinite reality).[7] Understanding all physical phenomena as expressions of matter and movement was essential to describing the universe. For Echegaray, heat, light, sound, gravity, and other physical manifestations could be related to movement:

> El calórico, según la teoría moderna, no es ya un nuevo cuerpo, un nuevo fluido, especie de sutilísimo gas que á manera de emanación va de una á otra parte, y donde se acumula produce calor, y desprendiéndose produce frío. El calórico, como la luz, como el sonido, como las olas del mar, como el astro que gira en el espacio en órbitas colosales, es MATERIA EN MOVIMIENTO; y así todos estos hechos se explican por la misma teoría, y están comprendidos dentro de una misma fórmula dinámica, que es la ley de la unidad—la gran unidad de todos estos fenómenos.[8]

(The caloric, according to modern theories, is not a new body, a new fluid, or a kind of light gas that, like an emanation, moves from one point to the other, and when it accumulates produces heat, or if it dissipates produces cold. The caloric, like light, sound, ocean waves, or a celestial body that revolves in colossal orbits in space, is MATTER IN MOVEMENT; and as such, all these occurrences can be explained with a single theory, and are described by the same formula of dynamics, which is the unity law—the great unity of these phenomena.)

These preliminary explanations, in which Echegaray offers a rational classification of physical phenomena, facilitate the detailed description of the theories and experimental results that appear later in the text. They are basic notions to which the reader could resort at any point to understand the particular depiction of society put forward in the sequel of the text.

Published ten years after the editorial success of the first part, this second volume uses a more specific approach to describing recent scientific developments in connection with social transformation and political change. Including specific images comparing the operation of society to an industrial process, the author not only offers simple and clear explanations of how many key concepts of physics work but also renders an interesting assessment of the country's social problems and the different tensions caused by modernization. Good examples of this are his description of the principle of energy conservation and his explanation of the mechanisms for transforming energy into work:

¿Veis al obrero levantar sobre el yunque la enorme maza, ó á esos otros obreros del pensamiento y del arte deslizar la pluma sobre unas cuantas cuartillas, ó apoyar el pincel sobre el lienzo, ó agitar con articulados sonidos el aire, escribiendo en sus fugaces palpitaciones las ideas? Pues todos esos movimientos, *como movimientos materiales*, eran, un año ha, vago bullir de unas cuantas moléculas perdidas en las inmensidades del astro de fuego. Después sus vibraciones se trasmitieron al éter, después bajaron en forma de luz, después repitióse en uno cualquiera de los vegetales de que el hombre se alimenta, ya directamente, ya por el intermedio de la carne animal, la pasada historia del carbón de piedra: y separado el carbono del oxígeno, fué sustancia orgánica y penetró en el cuerpo humano como el combustible en el hogar, y la corriente nerviosa le mandó arder, y ardió sin llama en el músculo, y engendró calor, y se convirtió en fuerza, y alzó el martillo, y agitó la pluma, y movió el pincel, y escribió palabras en el aire.[9]

(Do you see the worker lifting the enormous hammer over the anvil, or those other workers of the mind and the arts moving the pen over some pages or leaning the paint brush on the canvas, or making the air vibrate with articulated sounds, writing their ideas in brief palpitations? Well, all those

> movements, *as material movements*, one year ago were the vague vibration of some molecules lost within the immensity of our firing celestial body. After their vibrations were transmitted to the ether, they descended in the form of light, and then they were replicated in the vegetables from which people get their nourishment, either directly or through animal meat; mineral coal's old story: once oxygen and carbon separated, the last one became an organic substance that penetrated the organism, like the combustible does in the hearth, and then the nervous system commanded it to burn, and it burned without a flame in the muscles, generating heat and transforming it into the strength to lift the hammer and wave the pen or move the paint brush, and write words in the air.)

For Echegaray manual and intellectual labor were equivalent, in terms of both their importance for modernization and their implications for the forging of identity. In this vision of society, the articulation of culture depended on mutual collaboration between different social groups, all of which were essential constituencies of the national body.

As discussed in Chapter 1, in nineteenth-century technological imagination, humans, physically and mentally, were assimilated to machines. This analogy not only provided a particular understanding of the physical laws ruling energy, work, and movement, but also, and more importantly, of the human capacities for and responsibilities in invigorating modernization. In *Teorías modernas de la física*, Echegaray uses the internal functioning of the train locomotive as a metaphor for the body, the place where energy is transformed into human strength and reasoning: "El ser humano, como máquina, es una máquina de fuego, ni más ni menos que la locomotora. Compónese ésta, primero, de órganos en que se engendra la fuerza, como son el hogar en que arde el combustible, la caldera en que el agua recoge la vibración del fuego, el cilindro principal á cuyo émbolo trasmite al vapor la vibración recibida" (The human being, understood as a machine, is a combustion machine, no more, no less than a locomotive; it is composed, first, of organs where strength is created—such as the hearth, where combustibles burn, the boiler, in which water collects the vibration of fire, and the main cylinder, whose piston transmits to steam the vibration it receives).[10] Yet reducing humans to mechanical artifacts contravened the author's spiritual convictions—the human machine was a precise mechanism that transformed energy into productive work but only as long as it was animated by the soul, the metaphysical component that comprised human reasoning and feelings: "Queda, pues, explicado, y explicado brevemente, todo el mecanismo de la máquina humana; pero solo de la máquina, no del sér misterioso que siente y piensa y quiere" (Thus the entire mechanism of the human machine is easily explained; but only its mechanical part, not the mysterious being that feels, thinks, and loves).[11] This "mysterious" energy was precisely what differentiated body and mind—a Cartesian distinction that, in Echegaray's view, gave humans the possibility of conceiving abstract notions of the self and the collectivity, including concepts of affiliation such as those of national identity. By themselves, therefore, the joint material efforts of society were not enough to guarantee its progress; modernization would also require the synchronized spiritual energies of the country.

On the other hand, Echegaray's uses of the human body as a metaphor to explain the functioning of physics, particularly the principles of thermodynamics and the rudiments of electromagnetism, pertinently balance scientific materialism and spiritualty. His essays compare inventions such as the telegraph or the streetlight networks, for example, with the nervous system, the organic structure that supports emotions and feelings and that was believed to be the place where the soul and reason resided. This allegorical exercise leads to reflections on society's hierarchical configuration: "El sistema nervioso es una doble red telegráfica: ciertos alambres trasmiten noticias de la circunferencia al centro, son los conductores de la sensación; otros mandan órdenes del centro á la circunferencia, son los conductores de la voluntad. Y dicho esto, pocas palabras bastan para exponer las teorías de los materialistas modernos, fundadas exclusivamente en la sensación" (The nervous system is a two-way telegraphic network: certain wires transmit news from the periphery to the center, those are the transmitters of sensation; other send orders from the center to the circumference, those are the transmitters of will. Having said that, few words are enough to present the modern materialists' theories, which are exclusively grounded on sensation).[12] By explaining the nervous system as an electrical grid with a center-periphery topology that mirrored the country's administrative structure, Echegaray justified the state's centralism as a model of modernity that not only favored technology but also placed rationality at the center of industrial society.

As pointed out earlier, for Echegaray the popularization of science was a tool for discussing the country's complicated process of modernization while educating audiences on its foundational physical principles. Notwithstanding this goal's evident materialism, his approach heeded the metaphysical and spiritual aspects of human rationality—modernization was a quest for the balanced use of corporeal and mental capacities for the benefit of the country:

> Si el sér humano es un conjunto de movimientos, téngase en cuenta que en el movimiento sólo hay trayectorias, velocidades, forma, espacio y tiempo; que por muchos movimientos y muchas velocidades y muchas trayectorias que se acumulen, por mucho que se aumente la cantidad, mientras no varíe la calidad, jamás hallaremos ni sombra, ni remedo, ni nada que remotamente se asemeje á este maravilloso fenómeno de la conciencia.[13]

> (If the human being is reduced to a group of movements, one needs to take into account that movement is based only on trajectories, speed, structure,

space, and time; even though there could be many different movements and speeds, and many trajectories can accumulate, even when quantity increases, if there is no change in quality, we won't ever find an emulation, a shadow or anything remotely similar to the marvelous phenomenon of consciousness.)

Echegaray tries to simplify the theoretical complexity of thermodynamics and electricity by making clear distinctions between society's qualitative and quantitative attributes. No matter how big a collective was, if the quality of the work produced by all its members lacked the organization and logic of reason, there was no real gain in having them merging their efforts and energy. This exhortation also suggests certain challenges: in its path toward modernization, for example, the country's efforts needed to be allocated to reforms that facilitated the local production of science and technology.

Physical forces, Echegaray adds, needed to be properly understood—learning about their nature and interactions was thus of utmost importance in the efforts to control reality and contribute to society's progress. Within this process of intellectual discovery and enlightenment, nonetheless, intangible values also played a central role: only a combination of rationality and spirituality could really offer a complete vision of society and reveal the country's potential. In the conclusion of the text Echegaray explains: "Pues resulta claro como la luz meridiana, [*sic*] que las potencias que engendran al sér que piensa, [*sic*] no son las que engendraban el movimiento; que éstas son impotentes para aquello; que la materia grosera del mundo físico no es la que explica el mundo moral; que ha sido necesario suponer algo nuevo para justificar la existencia del nuevo y extraño sér que se llama hombre" (Well, it is clear as the midday light that those powers able to create a sentient being are not the ones responsible for movement; these are incapable of doing so. Vulgar matter that constitutes the physical world cannot explain the moral world; one needs to assume that there is something different out there to justify the existence of the new and uncanny being we call men).[14] Echegaray positions humans as the most complex product of the social environment, a context ruled by the same physical laws that shaped the landscape or allowed trains to function. What made humanity distinctive among other products of nature was the ability to reason and to understand people's agency in transforming reality. Without entering into metaphysical explanations, these divulgation essays give prominence to the duality of body and soul, material and spiritual components that needed to be in harmony to consolidate national identity as an amalgam of tradition and progress. As I show in the next section, in their evaluations of the country other popularizers and journalists of the period would also attempt to make compatible religion and scientific discovery in order to find a balance between the vital and rational forces that drove modernization.

Emilia Pardo Bazán: Critical Issues for a Changing Society

In addition to her work as a writer, translator, and cultural critic, Emilia Pardo Bazán was an active journalist.[15] Well versed in multiple fields, she received numerous offers to contribute with opinion pieces to the most important newspapers and journals of her time. The essays she dedicated to discussing Émile Zola's take on naturalism, as well as the ones aimed at popularizing Darwin's ideas, for example, are crucial to understanding the complex dialogue between a Spanish identity rooted in religion and foreign ideologies and aesthetic trends anchored to scientific materialism. Later included in the edited volume *La cuestión palpitante* (The critical issue) (1883), these texts refute biological and social determinism to defend free will from the standpoint of Catholicism. On that note, Pardo Bazán looked back to the problem of evolutionary biology, contending that Darwinism did not belong to the group of proven scientific truths, and, as a consequence, it did not comply with the requirement for evidence that Zola's positivist approach dictated. For her, Darwinism was only a daring system founded on some true facts and principles, yet full of unwarranted hypotheses with no solid proof, even when numerous intellectuals in "far away" countries, such as England, Germany, or Russia, were trying hard to find one.[16] Through that opinion, she not only stressed her ideological detachment from positivism but also marked the distance, physical and ideological, that separated Spanish modernization from that of the rest of Europe.

Pardo Bazán's essays join other texts of the time that contested or rejected materialism (in particular its emphasis on science as the only way to truth) to defend traditional Catholic dogma.[17] In 1877, for example, the first translation into Spanish of Paul Janet's influential essay *El materialismo contemporáneo* (Contemporary materialism) (1866) was published. In the text, Janet analyzes evolution's teleological inconsistencies to support the Kantian principle of the ultimate goal of things.[18] In Spain this analysis underpinned many of the arguments deployed in defense of Catholicism—revealing, as well, the significance that spirituality had for authors and for their conceptualization of national identity. But even though Pardo Bazán's journalism was aligned with this conservative posture, her work also vindicated scientific and technological developments as essential components of national modernization. In what follows, I examine three pieces of popularization in which she approaches topics such as electricity and thermodynamics, analyzes Darwinism, and reviews the advantages offered by the railroad. By understanding her take on what she dubbed "critical issues," it is possible to see a different angle in the assessment of the country's problematic modernization. As could be expected, Pardo Bazán's social foundry was essentially guided by a defense of Catholicism and the endorsement of traditional values.

As early as 1876, Pardo Bazán was commissioned to write a series of articles about what had become the scientific hot topics of the time: the caloric, electricity, the nature of light, and the principles that ruled movement. In similar fashion to Echegaray's illustrative volume on physics discussed earlier, in a series of articles entitled "La ciencia amena" (Enjoyable science) Pardo Bazán aimed to make advanced scientific concepts accessible to wide audiences. For that purpose, she "conveys Secchi's, Faraday's, Thomson's, and Berthelot's theories through her own ingenious metaphors."[19] These images were key for keeping the reader's attention and producing a coherent explanation of metaphysical notions, including the extension of the universe or the nature of reality. While language was deemed essential to establishing scientific truth, these rhetorical devices constituted an attempt to show that science and religion were compatible. For Pardo Bazán, the materialistic aspect of science did not supersede the existence of spiritual harmony in the universe—materiality and spirituality were just separate, and in some instances unrelated, aspects of the real world. In fact, contemporary science did not need to be in conflict with religion; since it was a theory mostly based on speculation, evolution, for example, could not even be confronted with the Catholic dogma. It was precisely in order to shed light on this sort of issues that communicating science had become an essential task, one that not only required accessible and *pleasant* descriptions but also clear distinctions between knowledge and faith:

> Alimento la esperanza de que aun la parte del público que no suele leer cosas de ciencia, no se arrepentirá acaso de haberme acompañado en mi viage [*sic*] al través de la creación. ¿Quién sabe? Quizás estas sencillas conferencias ofrezcan tanto atractivo, al menos, como una novela de esas que pervierten el gusto literario, sin enriquecer la mente con un solo dato, ni rasgar en lo más mínimo el tupido velo de la ignorancia.[20]
>
> (I keep the hope that even the portion of the audience who does not usually read things related to science perhaps won't regret having had me as a companion in this trip through Creation. Who knows? Maybe these simple expositions will offer as much entertainment as, at least, one of those novels that pervert the literary taste without offering a single fact to enrich the mind or slightly dissolve the dense cloud of ignorance.)

Yet Pardo Bazán's intentions with these essays were modest; her goal was supported by the principle that popular science and bad literature could be somehow comparable, since both captivated the attention of those who lacked the "taste" for more complex readings. At that level, a revision of scientific advances would complement, not challenge, the well-known story of Creation.

A key element in "La ciencia amena" is the use of images to which audiences could easily relate. As discussed in previous chapters, metaphorical systems are intrinsically linked to the ways in which reality is assimilated. For Pardo Bazán, incorporating commonplaces, or *topoi*, to explain complex phenomena, such as the existence of the caloric or the dynamic nature of electricity and light, served various functions. Through familiar images, she was able to establish an informal relationship with the reader and create the illusion of a shared layperson's condition with regard to scientific knowledge. By presenting materials without the conventional and sometimes arrogant tone of scientific disquisitions, she thus favored the notion of having an ideological affinity with her audience. Furthermore, the use of metaphors related to everyday activities, such as clothing, sailing, or ornamentation, was evidence of the extent to which advanced theories and concepts had penetrated the collective imagination. In Pardo Bazán's opinion, science's symbolic wealth transcended the senses (in fact, in most cases the inner operation of scientific notions escaped human perception) and was in direct dialogue with the elevated insight of the human soul: "Perspectivas tan curiosas [en referencia al estudio de estrellas distantes] no recrean los ojos, pero dilatan y embelesan el alma: y el estudio de la luz las proporciona innumerables. Estas páginas son solo una dedada de miel en tan rico y colmado panal" (Such curious perspectives [referring to the idea of distance stars] not only entertain the sight but also expand and captivate the soul: the study of light offers a myriad of these opportunities. These pages are nothing but only a pinch of honey in a full and delightful honeycomb).[21] This spiritual interpretation facilitated the appropriation of scientific truths into an imagination ruled by the strictness of religious principles to which Pardo Bazán, and most of her readers, subscribed. Yet there is both criticism and hope in her explanations: the readers she was hoping to reach were the ones that possessed imagination and curiosity, as well as a hunger for knowledge that went beyond the physical and sensorial world. Instead of finding pleasure in bad literature, Pardo Bazán expected her audience to find gratification in learning, a difficult task if one considered the weaknesses of the educational system.

Aiming to dissipate the fear of science as an antireligious form of knowledge, in her essays Pardo Bazán suggests the idea that scientific methods, theories, and proofs were more than abstract representations of reality; they were also palpable demonstrations of the existence of God. It was precisely in the fascinating correlation among the different phenomena of nature—electricity and magnetism, light and heat—where God's perfection materialized. At the end of "La ciencia amena," she thus concludes,

> En el maravilloso enlace del eterno mecanismo, se nos ocurre espontáneamente la idea de un todo armónico y soñamos la nota, la estrofa, la sinfonía. Así desaparecen de la naturaleza el azar, lo arbitrario, las combinaciones fortuitas, y adquiere nuevo sentido la frase bíblica tantas veces repetida, de que los cielos narran la gloria de Dios, y el común concepto de que en la sublime sencillez de la máquina se reconoce la destreza del artífice.[22]

(In the marvelous interconnection that articulates the eternal mechanism, the idea of a harmonic whole comes to us spontaneously, and we imagine the note, the stanza, the symphony. Hence chance, arbitrariness, accidental combinations disappear from nature, and the repeated biblical expression that the heavens are witnesses of God's glory acquires new meanings; so does the common idea that in the sublime simplicity of the machine it is possible to see the dexterity of the maker.)

Here, order and harmony reaffirm the divine nature of the cosmos, constituting strong arguments for refuting the central notions of Darwinism, particularly the concept of evolution as the result of random mutations. The real threat of science to the values of national identity, therefore, rather lay in the possibility of questioning the origin and purpose of life. It would be in her extended criticism of Darwin, written less than a year later for the journal *La Ciencia Cristiana* under the title "Reflexiones científicas contra el Darwinismo," that this notion of order as the unifying principle of nature would become central to her proposal.

Darwin's ideas had a disquieting effect in Spain. Soon after the publication of the first translation into French of *On the Origin of Species* in 1862 (Spanish renditions of the text would come later, generally using the French text as reference), conservative sectors manifested their uneasiness with the marked determinism implied in the notion of natural selection. Similarly, misinterpretations of Darwin's theory such as those of human evolution sharing roots with the development of apes implied, when assumed factual, direct attacks on traditional religious views.[23] Among the many sources of anxiety that Darwin's ideas and *Darwinismo*—their appropriation into the Spanish cultural and social imagination—yielded, the beliefs that science was able to contest God's purposes and refute the notion of free will were particularly problematic. Pardo Bazán's main criticism, for example, is rooted in the apparent conflict between the mutability implicit in evolution and the Catholic notion of harmony. For her, it was impossible to conceive any scientific truth that opposed or contradicted the basic notion of teleological predestination: "Bien se deja entender que el órden [*sic*] y concierto de lo creado suponen objeto y finalidad, y si el transformismo niega la finalidad, es á costa de romper la trabazón de la ciencia" (It is clear that the order and harmony of what has been created entail specific goals and finality, and if evolution denies finality, it is at the expense of facilitating scientific development).[24] Pardo Bazán stood for a vision of science that adhered to her own (and most Spaniards') religious beliefs. Her interpretation of evolution was not aimed at invalidating science or the scientific method per se, but at unveiling the possible inconsistencies between its premises and what she termed *Christian science*—hence the importance of the journal's suggestive name, *La Ciencia Cristiana*, and of Pardo Bazán's title for her series, "Reflexiones científicas."

In Pardo Bazán's social foundry, the contrast between science and religion functions as a benchmark for criticizing progressivists' insistence on accepting, even praising, scientific theories without questioning their validity. The kind of ideological polarization generated by Darwin's ideas was in many respects exceptional; other scientific discoveries that aided in the consolidation of industry or promoted material modernization were popularized (see, for example, Echegaray's work discussed in the previous section) without the characteristic antagonism of "Reflexiones." To soothe this animadversion, Pardo Bazán's vehement defense of traditional principles is articulated through numerous counterexamples that demonstrate, among others, how progress was not inherent to the processes described in evolutionary theory. On the contrary, notions of civilization are depicted as inconsistent with Darwin's argument. Thus, in order to reassure her audience that their beliefs were safe from "the attack" of science, in the text she celebrates the problematic transformation of the country, showing that industrial modernization produced undesirable consequences for society. Spain's resistance to renovating its identity was, to be sure, a positive trait.

One of the scenarios in which Pardo Bazán's social foundry underscores the fallacy of material progress as the preeminent path to social improvement is that of the dysfunctional industrial society:

> Lo primero se observa que, en lo tocante al mejoramiento físico, nuestra moderna civilización produce efectos perniciosos. Amontonada la multitud en las grandes capitales, marchitas en flor, por las devoradoras exigencias de la industria, las nacientes generaciones; creciendo cada día las necesidades, y con las necesidades el precio de los artículos indispensables, resiéntese de todas estas causas la robustez y gallardía general, y morbosos gérmenes inficionan la sangre de las clases trabajadoras, comunicándose por inevitable contagio, á las que no lo son.[25]
>
> (First, and with regard to material improvement, our modern civilization produces nothing but damaging effects. With multitudes piled in the big capital cities, and new generations withered in their prime by industry's consuming demands; with needs growing every day, and with them the price of indispensable articles. In that scenario, general bravery and vigor weaken, and harmful germs invade the working-class blood, spreading, by inevitable contagion, to other social classes.)

The problems of a growing working class are presented here as the natural result of a deviant and dysfunctional civilization. Consequently, medical language is used to describe *social unrest* as a communicable disease; even worse, the germ of rebellion could spread to other social groups. Other texts studied in this book approach industry in similar terms; those authors' social foundries, in fact, complement Pardo Bazán's assessment of modernization—for it implied rejecting the essence of what it meant to be Spanish, the excessive materialism of industrial progress was one of the main obstacles to assimilating the country's transformation.

Paradoxically, as long as science was confined to the material realm, it did not represent a real threat to traditional values. As a matter of fact, scientific theories and principles were seldom stigmatized as dangerous. For Pardo Bazán, it was the holistic pretensions of Darwinism that really constituted a problem: "La gravitación no dejó nunca los dominios de la física para entrarse por los de la metafísica: el darwinismo, más ambicioso, todo lo invade y pretende explicarlo todo, la nebulosa y el planeta, la célula y el hombre, el organismo del infusorio y la organización social" (Gravitation never left the realm of physics to enter that of metaphysics: Darwinism, more ambitiously, invades everything and tries to explain it—the nebula and the planet, the cell and men, infusoria and social organization).[26] Darwinism's spiritual and metaphysical aspirations, however, were not adequate to make it an acceptable way of approaching national problems.

If materialism and spirituality kept their distinctive limits, science and religion could have a harmonious relationship. By limiting their scope to physical phenomena, for example, physics and chemistry attested to the glory of God and the greatness of nature. Darwinism, in turn, was dangerous in its aspiration to explain not only the biological but also the spiritual components of being. Scientific progress could thus have both positive and negative effects on society: positive because it channeled the efforts of human intellect, and negative because, in doing so, it eroded tradition and led to social disarray. As a matter of fact, one of the biggest fears of the changes prompted by industrialization was a re-enaction of the French social insurrection of the mid-nineteenth century, particularly the possible uprising of the working class. Emilio Huelin, another science popularizer of the time, eloquently summarizes this anxiety:

> Las semillas del materialismo dieron frutos en el reinado de la *Commune*. Aquella gente había oído que la palabra Dios nada significa; que el alma humana es una cosa supuesta y vieja que ya no figura en la ciencia moderna, la cual enseña, según los materialistas, que el pensar es una secreción del cerebro, y la voluntad una vibración del sistema nervioso. De ahí dedujeron muchos que desaparecía toda responsabilidad individual, y que los crímenes castigados por la sociedad no eran más que resultados necesarios de una organización defectuosa.[27]
>
> (The seeds of materialism gave fruit in the kingdom of the *Commune*. Those people had heard that the word God did not mean anything; that the human soul is just an old and alleged idea incompatible with modern science, which, according to the materialists, proves that thinking is just a secretion from the brain, and that the will constitutes just a vibration of the nervous system. From that standpoint, many assumed that individual responsibility would disappear,

and that the crimes punished by society were no more than the necessary results of a defective organization.)

According to this interpretation, the emergence, consolidation, and expansion of the working class—patently incentivized by scientific and technological materialism—could destabilize society. Both Huelin and Pardo Bazán projected the fear of modernization onto their conservative view of social distinction and hoped for this structure to resist industrial change. In other words, for them the country's success as a modern nation depended on political, economic, and cultural strategies that somehow embraced the future while preserving and protecting the past.

Perhaps more pertinent for this discussion, and equally framed by some of the contradictions that articulate Pardo Bazán's social foundry, are the articles and stories that she dedicated to reflecting on her own experiences with the railroad. Written between 1896 and 1916 in her periodic column for the Barcelona journal *La Ilustración Artística* (Arts illustrated), these short reflections also touch, among many other topics, on the role social class played within the negotiations of progress. The way in which train travel altered different articulations of national reality also provides clues about people's anxieties with regard to the future of the country. As a frequent train traveler, Pardo Bazán's critique was aimed at the effects of what she saw as the mechanization of daily life. Understanding the railroad development as a measurement of modernity, her main preoccupation was the ensuing disarticulation of social structures and the misappropriation of technology.

A brief review of Pardo Bazán's short story "Sud-Exprés" (South express train) (1902) is useful here to theoretically contextualize her take on the railroad. The story focuses on the impact train travel had on the ways of representing reality—confinement, to take one of the many aspects that characterized this experience, for example, triggered in passengers' imaginations a sort of fictional mechanism, a rhetoric of suspicion based on the observation of *others*. By highlighting the social (dis)organization of modernization, the story uses this behavior to reflect on many of the profound anxieties that technological change was causing in the country. Finally, the text questions the collective nature of train travel, depicting passenger cars as isolated spaces wherein social codes were continuously renegotiated. "Sud-Exprés" is built on the notion of the train as a site of social simulation, a place where class structures were reproduced and, at the same time, adapted to the material conditions of movement and displacement facilitated by technology.

Given the speed and the limited visual contact with the exterior, train passengers were forced to find new ways of spending time during their trip. While the perceptual relation established between the traveler and the landscape dissociated interior and exterior realities, this extra time to ruminate also altered the division between reality and fiction. Train travel was in that sense not only an invitation to observe but an incentive to create. This introspection also induced reflections about the national character. In "Sud-Exprés," structure and plot are linked to particular notions of observation and imagination. In an exercise of purposeful oversight, for example, the narrator leaves behind descriptions of the scenery outside to concentrate in her surroundings. Thus, while the exterior was a setting where "los labriegos, las hortelanas que guiaban el carricoche atestado de hortalizas, al ver cruzar el raudo convoy, experimentaban esa impresión peculiar, de envidia respetuosa, que infunde el espectáculo de lo inaccesible social" (farmers and gardeners who guided covered wagons filled with vegetables, when seeing the fast train crossing experienced that peculiar impression of respectful envy that the spectacle of the social unreachable inspires),[28] in the inside people sitting next to her stimulated the sharp curiosity of someone suddenly immersed in an unknown but fertile environment.

Considering the trip a "social spectacle" reserved only for passengers, the narrator follows the peculiar behavior of a woman who seems to be living a double life, being unfaithful to her husband with a traveler in a different car.[29] The young woman takes advantage of the conditions provided by the train and, to the dismayed realization of the narrator, successfully manages to be affectionate to both men without being caught. The multifaceted life of this character then functions as a metaphor for the material conditions of train travel, setting the basis for future reflections about the railroad's efficacy and convenience as a means of transportation. Similar to a train schedule, the complicated drama of betrayal, unfaithfulness, and deception in the story takes place in a fractured timeline that the reader cannot completely understand: "¿Duró mucho el terrible y peligroso abrazo? Tal vez un segundo, tal vez cinco minutos o más." (Did the risky and dangerous hug last too long? Maybe a second, maybe five minutes or more.)[30] Without clear time or geographical references, during the trip the notion of duration and the sense of reality is altered. In this context, an ambiguous space emerged wherein events gained their own truth value—in "Sud-Exprés" there are multiple allusions to how the car's particular setting produces a blurry division between reality and fiction. A good example of this is the apparent continuity between the account the narrator is telling and the fictional story she is reading seconds before the other passengers' behavior disrupts her: "[La pareja] se situó tan cerca de mí, que su cuchicheo, impidiéndome fijarme en lo que leía, fue causa de que cerrase la novela de Danilewsky y prefiriese ojear la realidad próxima—sin sospechar que en ella encontraría, en vez de idilio, los elementos de un drama oscuro" ([The couple] sat so close that their whispering forced me to close the Danilevsky novel I was reading to look at the near reality—without knowing that, in it, I would find, instead of romance, the elements of a dark drama).[31] Danilevsky's text works here as one of the possible levels of truth, one not less fictitious than the "near reality," that is, than the narration itself. Pardo

Bazán's allusion to this popular Russian writer of historical novels is not fortuitous. It is precisely in historical fiction where subjectivity and objectivity converge to create the impression of documentary rigor within the context of literature's imaginative nature. This ambiguity is evident at the end of "Sud-Exprés," as the narrator questions the unusual situation she is in: "¿Efecto de mi vista miope? ¿Efectos de la imaginación? Hubiese jurado que era verdad" (Was it the effect of my shortsighted vision? Was it the effect of my imagination? I could swear it was happening).[32] While confinement altered the perception of time and the notion of truth, other conditions in the passenger compartment also opened the space for transgressing moral codes and social barriers otherwise consequential for the regular operation of society.

On the train, the unprecedented opportunity of being in close contact with *others* generated the idea of a fraternity of equals, a mutual understanding among people from diverse social and cultural backgrounds. This notion of collectivity, however, was problematized in the exterior, where an observer did not have access to the space of social interaction inside the train. For Pardo Bazán this separation mirrored class tensions, highlighting the sense of privilege and luxury usually associated with traveling: "Era una visión de cinematógrafo, desvanecida al punto mismo entre el penacho de humo y perdido en la distancia; y el hecho vulgar, sencillo, de almorzar así, servidos por camareros correctos, adquiría ante los espectadores, gracias a la velocidad del tren, a lo instantáneo de la imagen, una grandiosidad de alta vida, un realce novelesco y aristocrático" (It was a cinematographic vision,[33] faded between the crest of smoke and the distance; and the vulgar, simple fact of having lunch like that, served by proper waiters, acquired from the observer's perspective, thanks to the speed of the train, and to the instantaneous nature of the image, the grandeur of the high life, a fantastic and aristocratic splendor).[34] Story, novel (Danilevsky), and film are thus part of a metafictional construction that allows the narrator to be simultaneously outside and inside the train, placing her in a privileged position from which she can assess reality. This perception, however, was temporary, as was the idea of collectivity or the transgressive feeling of freedom created inside the train. Although "Sud-Exprés" explores movement and speed as creative engines, turning the invasion of privacy into a narrative exercise for the enjoyment of the readers, in reality Pardo Bazán was uneasy about the particular dynamics of socialization that the railway facilitated. In fact, as stated in many of her reviews of the automobile, she preferred other modes of transportation in which conservative notions of class stratification prevailed over the democratizing principles of progress.

Before becoming a car enthusiast, however, Pardo Bazán was already a seasoned train user; she wrote extensively about her experiences with the train in reviews and opinion pieces, articles in which she also evaluates the pitfalls and advantages of modern life. Aware of the country's social, political, and economic challenges vis-à-vis its resistance to change, the author resorted to journalism to denounce the backwardness of the country and reveal the different aspects of the Spanish character that, in her opinion, complicated national modernization. An ambivalent take on the country's development, this exercise reveals Pardo Bazán's particular understanding of this moment of transition—like other *fin-de-siglo* thinkers, she made a desperate effort to reconcile traditional values and progressivist ideas, which resulted in an eclectic assessment of reality.[35] An example of this wavery view is her argument favoring the railroad as the measure of civilization:

> En nada se refleja tan claramente la estrechez de nuestra vida moderna como en el corto número de trenes y su enlace dificultoso. Al acercarse á regiones donde hay vida industrial y fabril, Cataluña, Vizcaya, las pulsaciones de la circulación se acentúan, los trenes salen con frecuencia, el viaje se facilita y arregla de suyo. Pero donde la industria no ha exhalado su soplo bienhechor, los trenes van á paso de tortuga y salen con desesperantes intervalos.[36]
>
> (Nothing reflects more clearly the narrowness of our modern life than the low number of trains and their complicated system of connections. Approaching regions where there is some industrial and manufacturing activity, Catalonia, Biscay, the beating of circulation gets stronger, trains leave with frequency, traveling gets easier and more convenient. But where industry has not yet exhaled its beneficial breath, trains go at a snail's pace and depart in exasperating intervals.)

Spain's backwardness was therefore reflected in the precarity of its railroad system. However, by associating industrial operation with a circulatory system, Pardo Bazán's assessment was not entirely negative—in fact, the country is depicted as an active living organism with growing and dynamic potential, a body in which modernization could be the heart of that new and more dynamic Spain that the author longed for.

As pointed out earlier, industrialization altered the way in which time, distance, and space were understood. Those changes in the perception of reality led to the emergence of systems of representation that in some ways resonated with industry's new economies of production. Allusions to the railroad activity in relation to levels of *civilization*, such as the one included in the text just quoted, highlight precisely the importance of speed and effectiveness as elements that contributed to the "beneficial breath" of modernization. Pardo Bazán was aware of the complex interaction between the materiality of progress and daily life. Reduction in travel times not only improved productivity but also facilitated national cohesion by shortening distances and dissolving imagined borders—technology nullified, in many ways, the "narrowness" of Spanish social and economic development. As a consequence, the interactions of displacement, dynamism, and modernization not only resignified the notion of speed as an emblem of material progress but also changed the different forms of conceiving national identity.

Pardo Bazán's social foundry called for higher levels of symbiosis between people and the railroad; yet this interaction would both expose the system's greatest advantages and reveal its multiple risks. In any case, such information was essential to devising strategies to reduce the incompatibilities between the national character and technology. Pardo Bazán was perplexed by the way in which the many vices of the state administration had altered people's acclimation to the railroad. Instead of being an image of technological development and the progressivist spirit of nations, the system's late expansion and its poor management on the peninsula did nothing but confirm negative stereotypes of Spain as an exotic, almost barbarian place. In this perspective, uneasiness with the state administration was then translated into a negative review of the country's transportation infrastructure: "No debemos olvidar ni un segundo que los medios de comunicación son: (1) el camino de Europa, (2) la primera impresión por la cual Europa nos juzga" (Not even for a second should we forget that communication media are (1) the path that connects us with Europe and (2) the first impression Europeans use to pass judgment on us).[37] This emphatic depiction of the railroad as a bridge between cultures—as the link between Spain and Europe—aimed to call attention, first, to the geographical and imagined separation of the country from the rest of the continent, and second, to Spain's exceptional condition with regard to the standards of European modernization.

One recurrent topic in Pardo Bazán's articles in *La Ilustración* is Spaniards' lack of punctuality. The expansion and interconnection of railroad networks across Europe yielded the normalization of time zones and the inception of detailed train schedules.[38] The use of these essential standards in Spain, however, was not strictly enforced, generating dissatisfaction and annoyance among travelers. In her critique of this issue, Pardo Bazán assumes the role of a spokesperson, representing passengers and openly complaining about the indifference of Spanish authorities to comply with international regulations:

> No falta quien crea que si en España llega a desarrollarse cierta actividad industrial, y el sentido de los negocios se impone, se difundirá la perniciosa idea de que el tiempo tiene su valor y de que en todas partes el retraso de los trenes, salvo en casos excepcionalísimos y justificados, se castiga con multa y puede dar lugar á indemnizaciones. Pero esto será *ad kalendas græcas*, porque la piel del león de nuestro escudo hace rato que oculta á una tortuga entre sus crines.[39]

> (The only thing missing from this picture is someone thinking that if Spain achieves a certain point of industrial development, and a business mentality becomes the rule, this *harmful* idea that time is valuable would become

popular, revealing that everywhere in the world the delay of trains, except in extremely unusual and justified cases, is punished with fines and leads to compensation for passengers. But all this will happen *in a month of Sundays*, because it has been a while since the lion in our coat of arms hides a turtle underneath his mane.)

By suggesting that Spaniards did not have a clear notion of the cost of time, Pardo Bazán could only hope that stricter regulations would be in effect soon. Associations of time with productivity, which in other countries implied a connection between efficiency and profitability, are presented here as incompatible with local idiosyncrasies. Hence the author's allusion to the monarchy's coat of arms—wherein a lion represents the king's strength—as an emblem of an identity rooted in the mythical idealization of the past. This was, to be sure, an inaccurate reflection of reality, for in her critique it is a turtle that denoted the actual extension of the country's power. By superimposing these two images—turtle and lion, passivity and dynamism—past and present are contrasted, thus depicting a pessimistic image of a nation in which the benefits of progress had lost their value, and the train had become a referent of a deficient modernization.

Pardo Bazán's emphasis on productivity, efficiency, and velocity in relation to the train explains in part her fascination with the advent of the automobile. Scientific and technological developments in the field of electricity and combustion converged in this advancement that was supposed to leave behind the steam-engine era and with it all the ideas of slowness and backwardness that perfectly described the Spanish railroad system. By the second decade of the twentieth century, Pardo Bazán had already become a frequent user of the car and had begun to look at train travel with certain nostalgia—if speed was an indication of progress, some assertions about the automobile would contradict her initial criticism of the train:

Hay una manera de evitar el tren y sus molestias, que no son flojas: este medio es hacer el viaje en automóvil. . . . Por mí, lo juzgo el más grato. Nunca he dado suma importancia a la rapidez de los viajes: claro es que, en automóvil, como se ha de tomar algún descanso, no se irá tan aprisa como en tren, contando además con que el que en automóvil quiere ir aprisa, lo consigue y se hace polvo.[40]

(There is a way to avoid the train and its pains, which are not few: this solution is to make the trip by automobile. . . . For me, it is more agreeable. I have never given importance to the speed of traveling: it is evident that, in a car, as the driver has to rest, you cannot go as fast as in a train, not counting that in an automobile whoever wants to go fast can do it, but it may get killed trying.)

The "pains" of train travel are contrasted here with the advantages of not having time restrictions, or, in other words, the privilege of belonging to a certain social class in which punctuality and productivity were not determining or consequential. This ambivalence about the railroad—at the same time that the lack of punctuality is negative, the train's strict schedules can constitute a burden—is characteristic of Pardo Bazán's take on technology. Similar contradictions are evident in different aspects of her critique discussed here, including her classist tone or the reactionary perspective she deploys with regard to social divisions.

This equivocal posture is even more marked in her analyses of the issue of privacy and personal boundaries in trains. As noted in the analysis of "Sud-Exprés," train compartments generated spaces wherein the limits of public and private life blurred. The author's praise for the automobile emerges in part as the result of her uneasiness with this social reorganization:

> Entre las contingencias de la cabina, he omitido la de encontrarse en la mayor intimidad posible con una persona a quien no se conoce, y que en uso de su derecho la comparte con otra. Esto ya pasa de la raya, en cuanto a molestia. Calculo que aquel a quien le suceda, mal podrá conciliar el sueño. No es asunto de moral, pues hay cabinas para señoras, y en tal respecto no existe riesgo alguno; pero, descartado este aspecto de la cuestión, siempre quedará el otro: o no se aprovecha la cama, o se respira el aliento y se está en íntima conexión con gente desconocida.[41]
>
> (Among the possible eventualities in the compartment, I have omitted that of finding yourself in the most intimate space with someone you do not know, and who consensually decides to share that privacy with others. This pain oversteps the limits of what is admissible. I estimate that the one who suffers this contingency would have trouble getting to sleep. It is not a moral matter, for there are compartments only for women, and in that respect there is no risk at all; but, disregarding this aspect of the issue, others remain: either you don't enjoy the bed or one respires others' breath, ending in intimate connection with strangers.)

Pardo Bazán's preoccupation with the lack of boundaries, however, was not limited to the compartment; in the train station, as she notes in her reviews, the progressivist ideal of an inclusive society materialized, and all passengers, notwithstanding their social status, seemed to have access to the benefits of modernization. Yet this configuration generated discomfort in certain sectors of society who perceived these changes to contravene traditional class structures.

This contradictory appraisal of modernization respecting social difference is evident in the account of an incident she had at the ticket booth. After missing her train because of bureaucracy, Pardo Bazán wondered what would have been different had she established her status as a celebrity: "En el camino me explicaron que la carrera de obstáculos que encontré á mi paso era debida a que en la estación ignoraban *que yo era yo*" (On my way, it was explained to me that the obstacle course I had to endure was the result of the station's operators not knowing that *I was who I am*). Not happy with the outcome, she goes on with her rant: "Hablando en serio, ¿qué les parece á ustedes? ¿Verdad que la igualdad ante la taquilla debería ser un hecho? Porque, en la taquilla, esta igualdad existe ya en forma económica: todo el mundo paga—¡vaya si paga!—no siendo ciertos señores á quienes las Compañías llevan gratis y con sahumerio." (Seriously speaking, what do you think? Should not equality at the ticket booth be a fact? Because at the booth this equity already exists in economic terms: everybody pays—I bet they do!—unless they are the kind of people the company carries for free and with their incense burners).[42] Exposing the artificial nature of technological democratization, this assessment ratified the various misappropriations of progress in Spain, revealing the potential problems that the national character could imprint on the process of modernization. Although it was essential to make audiences aware of the need to align Spanish modernity with that of other countries in Europe, her analysis shows a hesitant and ambivalent attitude toward social change that seems to favor, for example, the less-collective experience of traveling by automobile.

Pardo Bazán's ideas on science and technology underwent multiple changes as society adopted and assimilated industrial transformation. The fast expansion of the railroad networks and services, for example, made traveling easier and faster; elements such as speed and displacement became common referents in understanding social change. It is in this context that the station, the compartment, or the train functioned as laboratories in which the country's problems could be diagnosed and prescribed. While modernization, in general terms, promoted the reconceptualization of national identity in relation to the past, the railroad's expansion was crucial to envisioning Spain as a modern nation. As discussed in the next section, however, the persistence of outdated social and economic models in the present would condition this projection.

Pedro Antonio de Alarcón: Roundtrip to Spain

Fascinated with the way in which train travel reconfigured people's perceptions of time and distance, many writers undertook the work of describing in great detail their encounters with the speed, their subjection to strict schedules, and their amazement at completing in a few hours a journey that in the past took several days. A bibliographical review of the period reveals that the most renowned writers of the second part of the nineteenth century documented in chronicles or reviews their expeditions. Works such as *Recuerdos de viaje por Francia y Bélgica* (Memories of a trip through France and Belgium) (1841) by Ramón de Mesonero Romanos; *Viajes al Vesubio* (Excursions to Mount Vesuvius) (1844) by Ángel Saavedra, Duque de Rivas; *De vuelta a Italia* (Journey to Italy) (1888) by Benito Pérez Galdós; *Cartas finlandesas* (Finnish letters) (1898) by Ángel

Ganivet; and *Por tierras de Portugal y de España* (Around towns in Portugal and Spain) (1911) by Miguel de Unamuno are just a handful of examples of this trend in Spain. During the nineteenth century, as William Stowe highlights, authors used traveling as a form of confirming their condition as intellectuals and social analysts: "It was a kind of work, and as such it could turn an idle holiday into a serious mission, a desultory grand tour into a purposeful investigation of foreign [or local] institutions, manners, and art."[43] Both fictional (novels, poems, short stories) and journalistic (chronicles, reports, letters) accounts of these excursions involved valuable assessments of society that were conditioned by the materiality of transportation technologies. Transcending the formal aspects of travel narratives, my aim in this section is then to explore elements such as speed, movement, or isolation to see how they establish the cultural parameters that defined certain social foundries.

In its most abstract aspect, train travel is bounded by two moments: departure and arrival. These two chronological markers, as Michel de Certeau proposes, designate a particular framework that is also constrained by the notions of interior and exterior.[44] As my analysis of Pardo Bazán's articles in the previous section showed, the physical conditions of isolation inside the train not only distorted the perception of time and distance but also established temporary codes of social behavior. Disconnection between the interior and the exterior of the train, for instance, contributed to feelings of social distance, thus changing passengers' interactions. In fact, the spontaneous parameters of socialization that emerged inside the train followed what George Simmel defines as sociological principles of segregation and homogenization: "A further quality of space, which has a fundamental effect on social interactions, lies in the fact that for our practical use space is divided into pieces which are considered units and are framed by boundaries—both as a cause and an effect of the division."[45] Isolating the interior of the train while keeping it integrated through a set of social parameters, the carriage created a unique universe of experiences with a great interpretative value. Technology's ability to abolish distance (through speed), moreover, transformed the interaction between the country's center and its periphery, as expressed by a popular saying of the time, "Lo que nunca se vio, el comer en Barcelona y cenar en Mataró" (This was never seen before: having lunch in Barcelona and dinner in Mataró).[46] Fast and effective communication also helped administrative authorities in their knowledge and understanding of the national territory, its cultural diversity, and its political dissonances. Travel accounts were in that regard essential for creating common ideas about progress and presenting different diagnoses of the country's situation. Additionally, since they were rooted in personal anecdotes, these testimonies depended on the authors' particular ideological vision.

Train travel fostered at least two key dynamics of socialization. First, travel's temporal nature created a sense of mobility, of *not belonging*, that facilitated encounters and interactions impossible in other contexts. Second, as already noted, physical proximity among passengers produced a false perception of recognition of the *others*, projecting a sense of *belonging* to an imagined collectivity. These contradictory perceptions had evident repercussions on the discursive representations of the travel experience itself, as well as on the authors' social foundries. Following these parameters, Pedro Antonio de Alarcón's *Viajes por España* (1883) exposes the insurmountable incompatibility between national identity and industrial modernization. Alarcón takes advantage of the spatial and temporal dislocation produced by the train and translates it into a juxtaposition of historical periods to highlight the glory of the past while acknowledging the social challenges of the present. The account of his trips around Spain is thus a *vehicle* to expose the country's attachments to tradition and its difficulties in embracing the advantages of progress.

Although it was compiled and published in 1883, *Viajes* reports on excursions undertaken by the author between 1850 and 1860, trips in which he rediscovered and explored a vast and practically unknown country. These journeys, which he qualifies as "expediciones artísticas o poéticas" (artistic and poetic expeditions), allowed him to see and understand many of the country's most urgent problems vis-à-vis the tensions between national identity and modernization.[47] Alarcón was born in Granada in 1833 and died in Madrid in 1891; during his life, he witnessed or participated in some of the most critical historical events of the period. Due in part to the political instability of those decades, his ideological posture regarding the country's social transformation greatly varied, going from a liberal and progressivist view in his first years in Madrid, where he arrived in 1854 escaping the reactionary environment of his native Granada, to that of a moderate conservative by the end of his life. Critics have paid particular attention to the moral aspects of this ideological transformation.[48] Cristina Viñes, for example, contends that Alarcón was the product of his own time and environment, a condition clearly visible in his professional trajectory.[49] This transition is also evident in his literary production, in which stories based on popular customs gave way to a realist and naturalist narrative consistent with the fin-de-siglo literary tendencies.

In addition to his prolific production in genres like the novel or the short story, Alarcón also wrote a substantial number of travel accounts. Most of those narrations originated in his work as a reporter for various newspapers. Because they were written over extended periods of time (even decades in some cases) before they were edited and compiled in volumes, differences in the style, structure, and content of the chronicles reflect the author's own ideological evolution. Excellent examples of his ability to depict historical reality and combine it with a particular political agenda can be found in anecdotical volumes like *Diario de un testigo de la Guerra de África* (Diary of a witness of the war in Africa) (1859), *De Madrid a Nápoles* (From Madrid to Naples) (1861), or *La*

Alpujarra: sesenta leguas a caballo precedidas de seis en diligencia (La Alpujarra: sixty leagues on horse preceded by six in carriage) (1873). Within this context, the interest of *Viajes por España* lies precisely in its self-referentiality, that is, in the way in which it approaches the problems of the country using Alarcón's own understanding of Spain.

Technology plays an essential role in many of the travel accounts included in *Viajes*. Elements such as the train speed, for example, are essential to describing the country's irregular but unique process of modernization. In the conclusion of the text Alarcón states, "He realizado muchísimas correrías, más o menos poéticas, por esta bendita tierra de España, donde me cupo la honra de nacer, y donde, dicho sea entre paréntesis, protesto vivir y morir *a uso y estilo* de mis difuntos padres, aunque cada día se invente un nuevo Paraíso terrenal al otro lado de los Pirineos" (I have completed many more or less poetic escapades in this blessed land of Spain, where I had the honor of having been born, and where, declaring it as a side note, I will live and die under the *same customs and manners* of my late parents, even when every day new paradises are conceived at the other side of the Pyrenees).[50] Emphasizing the complex assimilation of industrial social destabilization, the accounts included in the text offer detailed historical reconstructions that serve as guides to understanding the present. Descriptions of traveling on horseback, for example, are compared to the displacement by train to create an allegorical backdrop wherein the past, instead of being rejected, is refashioned to accommodate the changes of modernization.

Alarcón shows how the railroad, in spite of its novelty, was not completely dissociated from previous implementations; on the contrary, its lines had been built as a sort of palimpsest of older transportation routes, thus being not only a refashioning of tradition but also a reminder of the past: "De Valladolid a Palencia hay nueve leguas. Corren paralelamente a este trayecto la carretera, el canal de Castilla, el ferro-carril de Isabel II, el Telégrafo eléctrico y el río Pisuerga. Estas cinco vías se acercan unas a otras hasta el punto de hallarse unidas en algunos sitios dentro de cien varas de anchura" (The distance from Valladolid to Palencia is nine leagues. The road, the Castilian Canal, the Isabel II railroad, the electric telegraph, and the Pisuerga river run parallel to this path. At some points, these five channels of communication are as close together as to be in a one-hundred-yard space).[51] In Alarcón's depiction, this superposition of old and new means of transportation transcends the geographical description to become a reflection on tradition, and, consequently, on national identity. This perspective is particularly fruitful and informative when reading two of the accounts included in *Viajes*—his visit to the monastery of Yuste and his trip to Salamanca.

Alarcón's account of his visit to Yuste is the juxtaposition of two experiences separated in time. On the one hand, there is the author's report of his own journey to visit the famous monastery, and on the other, a portrayal of a similar trip that emperor Carlos V completed three centuries earlier to initiate his retirement:

"Por aquel escabroso camino, en que sólo nos restaba que [*sic*] andar algunos pasos, llegó Carlos V a su final retiro el día 3 de febrero de 1557, y por el propio sendero pasó su cadáver, después de haber yacido allí algunos años, para ir a continuar su sueño eterno en el panteón de El Escorial" (Following this rough path, on February 3, 1557, Carlos V arrived at the place where he was going to retire, and after a few years there his remains also made the same route to rest eternally in El Escorial's mausoleum).[52] Alarcón's description of the coat of arms decorating the monastery's entrance is also revealing; the emblem summarized and even epitomized an organic notion of Spain: "Aquel Escudo, abrigado por las poderosas alas del águila de dos cabezas y encerrado entre las dos columnas de Hércules, con la leyenda de Plus ultra, comprende en sus cuarteles las armas de todos los Estados del augusto Monje" (The coat of arms, sheltered by the powerful wings of the two-headed eagle, and enclosed by Hercules's two columns, with the inscription Plus Ultra [Further beyond], in its quarters comprises all the states over which the august monk exercised his power).[53] Yet, in spite of his painstaking reconstruction, Alarcón continuously reminds the reader that he was not writing a report on the historical trip of Carlos V, but on his own experience traveling to Yuste.[54]

By revisiting Carlos V's expedition, nonetheless, Alarcón's trip to Yuste turns into a journey to the past. While the account details events from several centuries before, formally it resembles a train itinerary, structured by separate instances in which each stop functions as a historical marker: "Partió el 6 de octubre de Laredo para Medina de Pomar . . . prosiguió ya más contento a Burgos, donde llegó el 13 y permaneció hasta el 16. . . . Marchaba tan lentamente, que empleó cerca de seis días desde Burgos a Valladolid. . . . Con esto partió de Valladolid (4 de noviembre), con tiempo lluvioso y frío" (He left on October 6 from Laredo to arrive in Medina de Pomar . . . then he continues, happier, to Burgos, where he arrived on the 13th and stayed until the 16th. . . . He was traveling so slow that he spent almost six days traveling from Burgos to Valladolid. . . . After that, he left Valladolid (November 4), with cold and rainy weather).[55] The rhythm of the narration here is shaped by modern-day imagery (the railroad); so is Alarcón's assessment of the present day, an aspect that becomes more evident in the detailed report of his journey from Madrid to Salamanca.

In contrast to the visit to Yuste, Alarcón's account of his train trip between Madrid and Salamanca proposes a social foundry based on the present to censure the desacralizing nature of progress:

> Habíase por entonces abierto al público la última sección del Ferro-carril de Medina del Campo a Salamanca, lo cual quería decir, en términos metafóricos, que esta insigne y venerable ciudad, monumento conmemorativo de sí propia, acababa de ser *desamortizada* por el espíritu generalizador de nuestro siglo,

pasando de las manos muertas de la Historia o de la rutina, al libre dominio de la vertiginosa actividad moderna.[56]

(Back then, the last section of the railroad from Medina del Campo to Salamanca had opened to the public, which, in metaphorical terms, meant that this emblematic and venerable city, a commemorative monument of itself, had been *disentitled* by the generalizing spirit of our times, going from the death hands of history or routine, to the free domain of vertiginous modern activity.)

For Alarcón, Salamanca's uniqueness (its tradition) was slowly disappearing amid the insubstantial nature of industrial activity. At the same time, the regular liveliness of progress was also positive "vertiginous activity" that opposed tradition's "death hands." Yet, in this depiction, modernization implies an expropriation of the past (literarily, a "disentitlement"), a break with beliefs and customs that antagonized liberal ideologies. In figurative terms, the train and the railroad were bridges that connected national identity with technology. Alarcón's complex reaction to modernization, shifting from disapproval to expectation, thus problematizes his criticism of tradition as a burden ("death hands") and of progress as a homogenizing force ("generalizing spirit").

This hesitant approach also evinces the prevailing tension between rural, traditional Spain, and the modern country built on industrial and technological advancements. The railroad, for example, destabilized society by connecting disparate social and economic spaces associated to either tradition or progress. While technology disentitled the past, it also generated opportunities to reconsider identity in light of the present. From Alarcón's perspective, the university city was the historical testimony of a national glory based on tradition—Catholicism and the monarchy. But, at the same time, by being linked to Madrid via the railroad, it also offered hope of refashioning that greatness in the present:

¡Considerad que allí hubo concilios; que allí se reunieron Cortes; que allí se juzgó a los Templarios; que allí se establecieron preferentemente las Órdenes Militares y fundaron magníficos templos; que allí predicaron San Vicente Ferrer y San Juan de Sahagún; que allí residieron mucho tiempo Santa Teresa y San Ignacio de Loyola; que allí estudió y explicó Fr. Luis de León, y que allí estuvieron los reyes Ordoño I, Alfonso VII, Fernando II, Alfonso IX, Enrique II (antes y después de matar a su hermano), D. Juan I, D. Juan II, D. Enrique IV, los Reyes Católicos (no una, sino muchas veces), el emperador Carlos V, Felipe II, Felipe III, Felipe V, y D. Alfonso XII, que felizmente reina![57]

(Consider that [in Salamanca] there were councils; that the National Assembly convened there; that the Templars were tried there; that military orders and

> magnificent temples were created there; that San Vicente Ferrer and San Juan de Sahagún preached there; that Santa Teresa and San Ignacio de Loyola lived there for a long time; and that kings Ordoño I, Alfonso VII, Fernando II, Alfonso IX, Enrique II (before and after killing his brother), D. Juan I, D. Juan II, D. Enrique IV, the Catholic kings (not one, but many times), emperor Carlos V, Felipe II, Felipe III, Felipe V, and D. Alfonso XII, who happily rules now, were there.)

Here, the possibility of connecting the conventional dynamism of the city with the long tradition of dignified greatness that resided in the village produces a visible opposition between the archaic nature of the rural past and the urban modernity of the present.

The contrast between these two temporalities in the text also highlights multiple forms of perceiving industrial advancement—perspectives, ideas, and impressions that often present contradictory or even incompatible arguments. A good example of these contentions can be found in Alarcón's description of a herd of fighting bulls grazing near the railroad lines:

> Lo primero que vimos de Salamanca (mucho antes de divisarla a lo lejos) fue sus célebres toros, . . . los toros salamanquinos, de mil libras de peso y de formidables astas, plantados cerca de la vía y mirando el tren con más cólera que espanto.
>
> —¡Ah, facinerosos! (estuve por decirles). ¡Desde tiempo inmemorial habéis estado yendo a Madrid a asustarnos con esa fuerza y esos cuernos que Dios os ha dado! . . . ¡Ahora nos toca a los madrileños venir a Salamanca a asustaros a vosotros!—¿Por qué no probáis a luchar con esta locomotora?
>
> Los toros debieron de adivinar semejante desafío, y noticiosos, sin duda, del trágico fin de aquellos héroes y mártires de su misma especie que embistieron arrogantemente en las orillas del Jarama a los primeros trenes . . . nos volvieron la espalda con suma dignidad.[58]
>
> (The first thing we saw in Salamanca (long before recognizing the city in the distance) was its celebrated bulls, . . . bulls from Salamanca, with one thousand pounds of weight and formidable horns, standing close to the railroad line and looking at the train with more rage than fear.
>
> Oh, you wicked! (I was about to tell them). From immemorial times you have been coming to Madrid to scare us with that strength and those horns that God gave you! . . . Now it is the turn of people from Madrid to come to Salamanca and scare you! Why don't you try to fight this locomotive?
>
> The bulls might have guessed that such a challenge was coming and, of course, well-informed about those heroes and martyrs from their own species who arrogantly charged at the first trains on the banks of the Jarama river, they turned away with supreme dignity.)

Because it exemplifies the vigor and value of tradition, the bull has been symbolically associated with Spain. This image shows the confrontation of that lore with the cognitive facet of progress—like in a bullfight, where the battle between the fighter and the animal offers a re-enactment of the eternal struggle between rationality and brutality. In this case, tradition seems to be unable to defeat the energy of modernization. From "immemorial times," the bull had represented the essence of a national identity that was intrinsically linked to the political and administrative center of the country; however, in the present, modernization, coming from that same center, was threatening tradition with a "tragic ending." Thanks to its material possibilities, in this vision progress's "generalizing spirit" defeats the national specificity represented by the bull.

A recurrent idea in Alarcón's social foundry is that modernization could be measured in relation to the level of development of transportation means—a notion one can see in other authors studied in this chapter. Technological assimilation, on the other hand, was delayed due to people's resistance to accept the transformations of progress. Incorporation of the train into daily life, consequently, entailed a complex negotiation of cultural and ideological dissonances. Even when Alarcón emphasizes the railroad's multiple advantages, in his descriptions there is an evident sense of fear and distrust that reflects those incompatibilities:

> Los españoles tenemos pocos asuntos fuera de casa, y los que tenemos no nos interesan hasta el extremo de hacernos emprender largos viajes. Nuestra filosofía moruna, ascética, o como queráis llamarla, da de sí esta magnánima indiferencia, tan deplorada por economistas y políticos, y tan aplaudida por otra clase de pensadores que miran las cosas desde más alto. Viajan, sí, por mero placer, los elegantes y los fantaseadores, los bañistas de afición y los amantes de la naturaleza.[59]

> (We, Spaniards, have few matters away from home, and the ones we have are not as important as to make us undertake long trips. Our Moorish philosophy, ascetic if you like, produces this magnanimous indifference, one that is constantly lamented by politicians and economists; the same one that, on the other hand, is applauded by other thinkers who observe things from a higher point. Traveling for pleasure is a thing, no doubt, but only for the elegant and the dreamer, the amateur bather, or the lover of nature.)

The railroad not only eliminated distances in physical (between one point and other) and allegorical (between past and present; center and periphery) ways, it also appropriated and rearticulated social restrictions. In the end, as Alarcón remarks, few people could afford to travel "just for pleasure," thus confirming the possible distortions of progress in a society attached to its traditional modes of living.

Projecting a positive image of technology, on the other hand, Alarcón depicts the railroad as a collective project to which every member of society had an obligation to contribute, an idea that reappears later in his description of the trip between Madrid and Santander. There, for example, national modernization becomes a feasible endeavor, one that, to be sure, had been delayed, but for which Spain had enough human capital and strength. Using an implicit reader, the writer interpolates himself: "Pero me dirás:—¿Cuándo llegas a Santander, a la capital de la provincia, al término de tu anunciado viaje?" (But tell me: When are you going to arrive in Santander, in the capital of the province, at the end of your announced journey?). To which the narrator replies:

> Llegaré, amigo mío, cuando *acabemos* el trozo de ferro-carril de *Los Corrales* a *Torrelavega*, en que *trabajamos* sin descanso, por medio de apuestas y de profecías, todos los habitantes de este valle, desde la distinguida familia constructora (inglesa por más señas), hasta mi humilde persona, que ha clavado ya más de una escarpia asentando *rails*. . . . —Conque ten otra semana de paciencia.[60]
>
> (I will arrive, my friend, when *we finish* this section of the railroad line between *Los Corrales* and *Torrelavega*, a project in which *all the inhabitants* of this valley have been restlessly working, in spite of bets and prophecies; from the distinguished family in charge of the construction (an English family, to be sure) to, humbly, myself, who has already hammered more than one spike to secure the rails . . . so be patient one more week.)

By presenting modernization teleologically, like the final destination of a trip that Spaniards had begun but not yet completed, Alarcón highlights the country's expectations with regard to the material possibilities of modernization. At the same time, the use of the plural first person, which is emphasized in the original text through the use of italics, allows him to include the reader as an active and important part of this collective effort.

Alarcón's social foundry uses specific rhetorical devices that exploit many of the railroad's peculiarities. This aspect becomes even more evident when the narration, due to the extraordinary character of the events described, transcends its chronological linearity and historical perspective. In those cases, the text suggests an interesting reflection on the nature of time and its impact on the assessment and interpretation of social problems—with the sequential nature of reality altered, tensions between the part and the whole, between the description of the general context and the detailed depiction of society, materialize. As I showed in chapter 1, in 1858, during the inauguration of the first line crossing the Cantabrian mountain chain, the author was involved in a tragic accident. The situation, as Alarcón explains, opened the door to new narrative possibilities:

> Las descripciones leídas de otras desgracias; la muerte imprevista; el mundo que desaparece; la familia; los amigos; el natural arrepentimiento del viaje; las personas que nos esperan; la fiesta frustrada; el instinto que clama por la conservación; el alma que condensa todo su poder, todas sus facultades para el instante supremo, y que, despidiéndose de sí misma, se dice: "Aquí era la muerte . . ."; todo esto y mil nimiedades que no sé cómo caben en aquella situación extrema, mil ideas frívolas, unidas a otras muy solemnes y graves, la muleta, la mano cortada, lo que será uno sin dientes, la cuestión de la inmortalidad del alma, lo que dirá fulana cuando sepa lo sucedido, cómo llegará la noticia al hogar paterno, y un punto de conformidad cristiana, y una mirada al cielo, y la tranquilidad más estoica, y el miedo más miserable: todo eso y mucho más, resumido en una idea multiforme, súbita, luminosa, intuitiva, llenaron aquellos cuatro segundos, abreviatura y término de la existencia.[61]

> (The literary description of similar tragedies; the unexpected death; the world suddenly disappearing; family; friends; the natural remorse of traveling; the people who wait for us; the aborted celebration; the instinct of preservation; the soul condensing all its power, all the faculties reserved for the supreme moment, and saying goodbye to itself exclaims: "Here comes death . . ."; all this, and a thousand trifles more than it seems possible to fit together in that extreme situation, a thousand shallow ideas linked to many other very solemn and serious thoughts—the crutch, the severed arm, how one would look with no teeth, the question of the immortality of the soul, what Tom, Dick, and Harry would say about it when they learn of the tragedy; and a gaze to heaven, and a stoic calm, and the most miserable fear; all this, and much more, summarized in a multishaped idea, sudden, bright, intuitive, that filled those four seconds: the encapsulation and conclusion of existence.)

Resorting to "literary descriptions" of similar tragedies, the author found himself condensing his life into four seconds. The passage captures in images the burst of ideas, memories, and places that crossed his mind when he confronted death. Witnesses' impressions of the event complement the appropriation of elements, such as speed, destructive forces, and fear of the unknown; perceptions that can only be captured in the text with the use of a fragmented narrative resembling, as in the account of Carlos V's trip to Yuste, a train itinerary with multiple stops.

In other circumstances, however, the train's speed created a different perception of time—one related to the metaphorical fusion of past and present. Movement indeed yielded a sort of visual integration between the vehicle itself (a materialization of the possibilities of progress) and the surrounding areas crossed by it (usually rural spaces associated with the essence of tradition):

> Parecía aquello una sombra de ferro-carril . . . pero yo me alegré en el alma de hacer aquellas nueve leguas tan solitaria y cómodamente, corriendo de una ventanilla a otra para admirar soberbios paisajes montañosos, en que se veían confundidos árboles, rocas, malezas, viaductos, prados, cabañas, túneles, desmontes, bosques, arroyos, puentes. . . . ¡Todos los encantos de la naturaleza y de la civilización![62]

> (The system was more like a shadow, an early image of what was supposed to be a railroad . . . but I was happy in my soul at completing those nine leagues in such a lonely and comfortable way, running from one window to the other to see the superb mountainous landscape, in which trees, boulders, weeds, viaducts, lawns, cabins, tunnels, cleaned areas, woods, rivers, bridges fused. . . . All the charms of nature and civilization!)

Praise of technical advancement and admiration of nature contends with Alarcón's initial warning about modernization's ability to destroy history, an idea common to many of the accounts analyzed here. Yet it is precisely this kind of duality that made the appropriation of industrial elements a rich and complex process.

The significance of the descriptions included in *Viajes por España* lies in Alarcón's dual position as a participant and a distant observer of social change. This condition facilitates his use of technological imagery to codify different aspects of national reality and confront them with the challenges and achievements of society. As is shown in chapter 6 of this study, the experience of technology not only provoked reflections on national identity but also stimulated considerations of other social problems of the time. Environmental concerns, among other phenomena associated with the negative impact of modernization, emerged in this context as yet another telling referent in the evaluation of the country and the visualization of its future.

CHAPTER 6

Industrial Footprint

Industrial production in Spain was not always concentrated in cities: in fact, sectors associated with the mining and steel industries (two of the main sectors of the economy, as shown in previous chapters) settled outside metropolitan districts, thus subverting the dynamics of dependency between center and periphery in terms of wealth generation and knowledge exchange. An additional effect of this peculiar inversion was the exacerbation of disparities between rural and urban areas—first between the administrative capital and the provinces, and then between industrial centers (formerly small villages) and their surrounding areas. In the periphery, social structures resisted industry's transformative capacity. This reluctance complicated the assimilation of a modernizing force that was promptly associated with the metaphorical destruction of the past and the physical devastation of nature. Ecocriticism, as Timothy Clark suggests, is useful in this regard to evaluate the different factors—material, cultural, psychological, legal, political—at work in environmental degradation as the result of human intervention.[1] This chapter shows how rural ecosystems' reconfiguration in *fin-de-siglo* Spain changed writers' particular grasp of the country and its problems.

Under industry's new regimes of human and economic interaction, different social dynamics of regrouping, segregation, and alienation took place. Capital accumulation in a few hands, for example, altered class tensions by questioning traditional configurations of political and economic power; meanwhile, the continuous migratory displacement of populations progressively transformed the social codes and tacit agreements of coexistence in industrial settlements. As noted earlier, in this context the social-class consciousness that had endorsed the modernizing efforts in the country dissipated. Social mobility became simultaneously a risk and a challenge for the reformulation of a national identity that was expected to be compatible with the new conditions imposed by modernization.

This process of social reconfiguration exacerbated modernization's environmental issues—industry's negative effects on nature hindered the transition from rural-based structures to modern urban economies. The natural world, rather than being the realm of passive entities at human disposal, constitutes a dynamic network of interactions wherein people and things ideally work together to the advantage of both.[2] Technology altered this symbiosis, creating specific scenarios in which contradictory projections of the country were viable. As discussed earlier, discursive production was particularly susceptible to this ambivalence: in a single author it was possible to find feelings of admiration for the advancement of science and technology, as well as sentiments of fear and rejection of the effects these advances could have on society. A good example of the nature and extent of this equivocal reading of progress appears in Emilia Pardo Bazán's reflection on her visit to the 1889 World's Fair in Paris. Referring to the Eiffel Tower, a prominent emblem of the technological and industrial progress of the period, she states,

> La que llamamos civilización ¿es más que una batalla sin tregua para ganar un pan amargo, para cubrir necesidades ficticias y para vivir roído de cuidados y en ahogo perpetuo? Y cuando decimos que hemos llevado la luz, la ciencia y el progreso a una región salvaje, ¿no podríamos añadir que llevamos la inquietud, el desasosiego y las penas del alma? . . . He ahí lo que representa nuestra brillante civilización para el minero: sepultarse todos los días a 530 metros bajo el nivel de la superficie terrestre.[3]

> (Is that which we call civilization something different than an endless battle to earn a bitter piece of bread, to fulfill fictitious needs, and to live consumed by perpetual calamities and exhausting search for care? And if we say that men have brought the light of science and progress to a wild region, should we also add that men have brought unrest, feelings of uneasiness, and pain to the soul? . . . Hence, we will see what our brilliant civilization really means for a miner: burying himself every day 1,600 feet below ground level.)

Censoring society's incessant quest for modernization, Pardo Bazán questioned the alleged positive effects of technological development. In the passage, contrasts between light and darkness, for example, work as referents to confront civilization with its human and environmental cost: in order to raise a tower of nearly a thousand feet, reaching the highest point of technical development, workers needed to bury themselves a similar distance under the ground. Whereas the tower was a visible image of progress, the laborers' sacrifice and violence against nature underpinning it were hidden from the observer—two sides of modernization that highlighted irreconcilable aspects of progress.

In addressing tensions between past and present, religion and science, or rural and urban spaces, the literary works analyzed in this chapter reveal new aspects of the dynamic and unstable essence of Spanish modernization. The first section examines Benito Pérez Galdós's early novel *Marianela* (1878) to show his assessment of science as part of the solution to society's most urgent problems. I highlight the importance he gives to the idea of a modern Spain that needed to break links with the past, no matter the human or environmental cost, in order to confront the challenges of the present. The second section explores José Ortega Munilla's novel *El tren directo* (Express train) (1880). In this story about the disruptive capacity of progress, Ortega indirectly questions Galdós's assumptions about the importance of modernization, showing that industrialization and material development in general, more than constituting an opportunity, could also, and perhaps more importantly, be a threat to national identity. The final section deals in depth with the feelings of ambivalence that the social and environmental transformations of industry generated in the population. Perceived as a moral deterioration of national values, natural destruction functions here as a metaphor for a lost paradise. Armando Palacio Valdés's novel *La aldea perdida* (The lost village) (1903) aptly illustrates this fascinating aspect by contrasting the past and present of an idyllic rural society in which the arrival of workers destroys the harmony of daily life and limits the possibilities for progress.

Through a revision of the industrial footprint as depicted in these three novels, the chapter also shows how literature interacts with the social technologies of modernization to present unanticipated aspects of industry's epistemological shift in fin-de-siglo Spain. In contrast to social, political, and economic diagnoses explored in previous chapters, the multilayer textuality of literature deployed in these three novels offers new and unexplored possibilities in the articulation of *social foundries*.

Blindness, Reality, and Representation in Galdós's *Marianela*

Geographer Peter Taylor has identified three key processes in the consolidation of national identities: rediscovery, refashioning, and invention. For Taylor, ideas of nation and identity start with the rediscovery of local myths, which are later refashioned to fit the collective expectations of a particular historical moment.[4] Deployment of industrial images to propose an assessment of modernization, for example, is a common narrative strategy in many of Benito Pérez Galdós's novels that in one way or another deal with the problems of national identity. Galdós's vast oeuvre and its importance within the fin-de-siglo literary corpus have been widely studied. Traditionally, critics have approached his writings by tracing links between the country's social and political reality and the philosophical and aesthetic tendencies of the period.[5] The novel analyzed in this section, for example, has been understood as an example of the strong influence of positivism on the author's early work. Cultural critics, such as María Paz Yáñez, Joaquín Casalduero, or C. A. Jones, read *Marianela* as one piece in a bigger literary project dedicated to studying the effects of scientific rationalism on the consolidation of modern society.[6]

Marianela expands on ideas previously explored by the author—the ideological conflicts in *Doña Perfecta* (*Dona Perfecta*) (1876), religious tensions in *Gloria* (1877), or the materialist predicaments of characters in *La familia de León Roch* (*Leon Roch: A Romance*) (1878), for example, question the assimilation of industrial development in a contentious society wherein passion and personal interest seemed to prevail over reason and the search for the common good. In the text, contrasts between past and present highlight the difficult process of (re)shaping national identity, thereby exposing the negative impact of modernization on nature and, ultimately, on the structure of society. As recent scholarship on the novel aptly points out, the repercussions of industrial activity are depicted through different manifestations of violence—against the environment, the social body, or tradition.[7] Analyzing images that problematize society's attachment to the past within this context, in what follows I read *Marianela* as a critique of the unsettled possibilities of the country. In Galdós's social foundry, notwithstanding the elevated human or environmental cost, Spain needed to embrace the ideals of progress embedded in the liberal plans for social and political reform of the revolutionary period.

Galdós's affiliation with realism in his early production has been a matter of debate. For some critics, the presence of certain idealism in novels and stories published before 1880 tarnished the possibility of rendering an objective representation of reality. Texts from this period, more precisely, evince a subjective take on the country's religious and political tensions. In this regard, Juan López-Morillas has pointed to a certain "despotism of ideas" that, from his perspective, prevailed in most of the Spanish literature of the 1870s as the result of the strong idealism that surrounded the *Revolución Septembrina*. Only after 1880, López-Morillas explains, once Krausism's intellectual ebullition soothed, was the production of realist literature—that is, narrative depictions of reality stripped of any ideological conditioning—possible.[8] Yet it is precisely the strong idealism characteristic of Galdós's early work that turns *Marianela* into an illustrative reflection of the conflict between reality and representation.

Considering the illusory essence of progress, *Marianela* proposes an original exploration of the social and political instability of the period. In the story, Teodoro Golfín, a renowned ophthalmologist from Madrid, arrives in the mining village of Socartes to cure the blindness of Pablo Penáguilas, only heir of the most prominent family in town. The famed doctor's services had been requested through his own brother, Carlos Golfín, a chief engineer in the mining complex. The surgeon's arrival, however, destabilizes Pablo's world—an idealized reality he has conceived in his mind with the aid of his friend and guide, Marianela. Poor and unattractive, this young girl's aspiration is to someday become Pablo's wife. Unaware of the social restrictions that would prevent her from having her dream come true, a disillusioned Marianela ends up being rejected once Pablo's vision is restored. Not only did Marianela erroneously think that class differences were not impediments to her dreams, but she was also confident in the nature of Pablo's feelings toward her. In fact, she is appalled at Pablo's decision to marry his cousin Florentina, the embodiment of the bourgeois ideal of beauty. In the end, these developments lead Marianela, first, to attempt suicide, and finally to die from a broken heart. Charged with blame, the Penáguilas decide to adorn her tomb with excessive decor, turning the site into a tourist attraction that gives the visitors the impression that Marianela was an important member of Socartes's society and perhaps one of the most beautiful women in Spain. With a great sense of irony, the story highlights that illusion can be more powerful than reality.

Beyond the love story, however, *Marianela* is also a tale of industrial transformation and scientific success. One of the most powerful images projected by the text is that of the *triumphal science*, an optimistic vision of modernization in which knowledge and rationality could cure the blindness of society. As Andrew Anderson's recent analysis of the novel shows, this is indeed an eloquent image of progress: parallel to notions of treatment and disease, ophthalmology's positive capacity for transformation can easily be contrasted to the negative impact of mining.[9] In fact, medical discourse and scientific ideas work as allegorical frameworks in which society's ailments can be diagnosed, prescribed for, and eventually cured. With the purpose of highlighting the weaknesses of the past and the great possibilities of the present, Galdós focuses on describing Marianela's unhealthy complexion in contrast to the strength and physical beauty of other characters. This opposition is extended to other dichotomies articulated in the novel, such as blindness and sight, or natural chaos and industrial order. Moreover, scientific reasoning is at the center of the novel's narrative structure and, on different levels, there is clear evidence of the influence of positivism on Galdós's ideas—Jones, for example, has identified in the text the three stages of evolutionary development proposed by French philosopher Auguste Comte: "Marianela herself as representing imagination, Pablo as rationalism, and Teodoro as science."[10] Within this progression, excess of imagination was thus one of the main obstacles for the consolidation of a modern, rational Spain. Galdós's social foundry concurs with that of scientists and educators such as Echegaray or Mallada, who envisioned a national identity built on concrete reasoning instead of idealization. *Marianela* in effect focuses on showing the efforts of scientists (a surgeon, in this case) and engineers to control the physical and cognitive transformations of society.

The dynamic character of modernization is emphasized in the novel through the description of industry's destructive capacity.[11] Environmental damage is used as a metaphor for the radical and violent transformation of society. Golfín's first encounter with the mining landscape, at the beginning of the novel, for example, underlines progress as an ambiguous force—a positive agent of change and a negative instrument of destruction:

> El viajero, que había andado algunos pasos junto a su guía, se detuvo asombrado de la fantástica perspectiva que se ofrecía ante sus ojos. Hallábase en un lugar hondo, semejante al cráter de un volcán, de suelo irregular, de paredes más irregulares aún. En los bordes y en el centro de la enorme caldera, cuya magnitud era aumentada por el engañoso claro-oscuro de la noche, se elevaban figuras colosales, hombres disformes, monstruos volcados y patas arriba, brazos inmensos desperezándose, pies truncados, desparramadas figuras semejantes a las que forma el caprichoso andar de las nubes en el cielo; pero quietas, inmobles, endurecidas. Era su color el de las momias, un color terroso tirando a rojo; su actitud la del movimiento febril sorprendido y atajado por la muerte. Parecía la petrificación de una orgía de gigantescos demonios; y sus manotadas, [así como] los burlones movimientos de sus desproporcionadas cabezas habían quedado fijos como las inalterables actitudes de la escultura.[12]
>
> (The traveler, after taking some steps into the mine in the company of his guide, stopped and contemplated with perplexity the fantastic perspective offered to his eyes. He was in a deep location, similar to the crater in a volcano, with an irregular floor and even more irregular walls. On the edges and at the center of the enormous caldera, the real size of which was altered by the deceiving contrasts of the night light, colossal, deformed figures arose—overturned and upside-down monsters, huge arms stretching out, truncated feet, scattered figures similar to those formed by clouds when moving in the sky; but static, motionless, hardened. Their color was the same as that of mummies, an earthy color close to red; their attitude was one of a feverish movement surprised and stopped by death. The spectacle resembled the petrification of an orgy of giant demons, and the demons' big hands, [as well as] their disproportioned heads, had become fixed like the unchanging attitudes in a sculpture.)

The "giant demons" that populate this new landscape are part of the remnants of industrial activity. In them, a combination of science, art, and destruction is condensed—they are sculptures shaped by modernization's capacity for transformation. Golfín has difficulties in making sense of this peculiar landscape in which progress, imprinted on nature, simultaneously shows its positive and negative effects. Positive because it evidenced human ability to modify its surroundings; negative because it disfigured beauty and produced suffering. This image of a harmonious and beautiful scenery in which monstrosity is hidden encapsulates the tension between appearance and reality that articulates the novel.

Figurative uses of the body as a metaphor for industrial mining complement this physiological evaluation of reality. In Golfín's descriptions of the mine's galleries, for example, there is a marked medical tone that facilitates this diagnosis: "Este pasadizo es un exófago [*sic*]. Somos pobres bichos que hemos caído en el estómago de un gran insectívoro" (This gallery is an esophagus. We, people, are just bugs that have fallen in the stomach of a big insectivore).[13] Humans and nature are at industrialization's mercy, their weakness and fragility can only be compared to those of an insect unable to fight its predator. By using digestion as a metaphor, Galdós reaffirms the idea of progress as both agent of change and instrument of destruction; a similar image that the one deployed later to characterize medical science. Analogously, Marianela's physical appearance resists a definite assessment: "Alguien decía que era una mujer mirada con vidrio de disminución; alguno que era una niña con ojos y expresión de adolescente. No conociéndola, se dudaba si era un asombroso progreso o un deplorable atraso" (Some people claimed that she was a woman seen through a demagnifying glass; some said that she was a child with the eyes and expression of a teenager. Without knowing her, one hesitated between considering her a miracle of progress or a deplorable regress).[14] Giving particular importance to sight, Galdós thus put forward a new *way of seeing* both negative and positive connotations of tradition in the context of modernization.

Industrial images are also used in the text to illustrate and criticize different aspects of the environment. Descriptions of Socartes, for example, show the small village and its inhabitants as extensions or constitutive parts of the industrial apparatus:

> El humo de los hornos que durante toda la noche velaban respirando con bronco resoplido se plateó vagamente en sus espirales más remotas; apareció risueña claridad por los lejanos términos y detrás de los montes, y poco a poco fueron saliendo sucesivamente de la sombra los cerros que rodean a Socartes, los inmensos taludes de tierra rojiza, los negros edificios. La campana del establecimiento gritó con aguda voz: "al trabajo," y cien y cien hombres soñolientos salieron de las casas, cabañas, chozas y agujeros. Rechinaban los goznes de las puertas; de las cuadras salían pausadamente las mulas, dirigiéndose solas al abrevadero, y el establecimiento, que poco antes semejaba una mansión fúnebre alumbrada por la claridad infernal de los hornos, se animaba moviendo sus miles de brazos.[15]

> (Making a rough breathing noise, smoke coming from ovens that have worked all night turned into a gray spiral; the morning light appeared from behind the far mountains, and, little by little, the mountains surrounding Socartes—huge slopes of reddish ground similar to black buildings—became visible after night shade left. At that moment, the factory's bell yelled with acute voice, "To work!" and hundreds of drowsy men left their houses, shacks, and holes. Door hinges grinded; from the stables, mules slowly left by themselves heading in the direction of the drinking trough, and the factory, which earlier

> resembled a funeral house lit by the infernal clarity of the ovens, came to life moving its numerous arms.)

In this image industry has not only absorbed workers and animals, but also predated the landscape—buildings are thus black as coal, the same mineral that provides energy for the industrial operation and income for the population. It comes as no surprise that workers are depicted as pieces of mineral, pointing to the fact that they constitute industry's main combustible: "Hombres negros, que parecían el carbón humanado, se reunían en torno a los objetos de fuego que salían de las fraguas, y cogiéndolos con aquella prolongación incandescente de los dedos a quien llaman tenazas, los trabajaban. ¡Extraña escultura la que tiene por genio al fuego y por cincel al martillo!" (Black men, resembling humanized coal, gathered around the firing objects that the forge expelled, and holding them with that prolongation of the fingers called pliers, they manipulate them. What an unusual sculpture is made by the artistic genius of fire, chisel, and hammer!)[16] Here, images of work and energy perfectly capture the dialogue between industry and art, science and passion, rationality and spirituality that Galdós places at the center of his social foundry.

In the novel the tension between past and present (temporalities that mirror the aforementioned dichotomies) is also transposed onto the characters' moral constitution. Galdós criticizes the social toll of scientific conquest—Golfín's reservations about proceeding with Pablo's treatment in the face of the negative consequences it could have for Marianela is a good example of this. The cost of Pablo's *seeing* again is the loss of everything that is valuable for Marianela: imagination, religiosity, purity, freedom, tranquility. As representative of the liberal bourgeoise, however, Golfín's main goal is securing the country's future by eliminating the limitations that the *lack of vision* produced in Spain. Using the tension between idealization and reality as a referent, the doctor explains, "La realidad ha sido para él [Pablo] nueva vida, para ella [Marianela] ha sido dolor y asfixia, ha sido la humillación, la tristeza, el desaire, el dolor, los celos . . . ¡la muerte!" (Reality for him [Pablo] has been life; for her [Marianela], it has implied pain and asphyxiation, it has been humiliation, sadness, disgrace, pain, jealousy . . . Death!)[17] In regard to this contrast, Jones adds, "*Marianela* emerges as an experiment among others in the search for an answer to the problem of how a man can come to terms with reality."[18] Pablo's construction of reality as an extension of his imagination, and Marianela's idealized *vision* of the world perfectly articulate this narrative approach.[19]

Through allegorical confrontations between reality and imagination, Galdós also addresses the risks of glorifying modernization. Before recovering his vision, for example, Pablo's love for Marianela was conceived out of an idealization of beauty. Her profound knowledge of the terrain and dedicated care as a guide had allowed Pablo to confidently navigate the mine's complicated geography—thanks to Marianela, he was able to move through modernization even when he could not actually apprehend it. Yet Pablo had built an erroneous image of the world, an almost-platonic idealization of life.[20] He was convinced that Marianela, a low-class woman with physical limitations, could be an ideal match for forming a family. Once he *recovers vision*, the possibilities of materializing such a union, in which the rational and financial capacity of the bourgeoisie would join the proletariat's productive forces, vanishes. In banishing the possibility of mixing social classes, Galdós's critique was not aimed at exposing the prejudices of a conservative society, but rather at the high cost of the country's leaving its past behind to face the challenges of the present.

Galdós's use of science as a metaphor in the novel serves the purpose of highlighting the incompatibility between reason and passion. Golfín's hesitance to proceed with Pablo's surgery, in fact, exposes some of the consequences of his ethical and ideological conflict: "Problema y duda tenemos aquí. . . . Pero hagámosle hombre; ese es el deber de la ciencia; traigámosle [a Pablo] del mundo de las ilusiones á la esfera de la realidad, y entonces, dado su poderoso pensar, será verdaderamente inteligente y discreto; entonces sus ideas serán exactas y tendrá el don precioso de apreciar en su verdadero valor todas las cosas" (Here, we have a problem and a doubt . . . but let us proceed; such is science's obligation; let us bring him [Pablo] from the world of illusions to the sphere of reality, and then, given his powerful mind, he will be truly smart and discrete; then, his ideas will be exact and he will possess the precious gift of valuing things for their real cost).[21] For the doctor (perhaps a fictional alter ego of Galdós), rational knowledge was more valuable than imagination and poetry. If all this was true, then it could be said that Pablo had been following the wrong guide to navigate industrial reality. Golfín's description of Marianela further elaborates on this point: "Posee una fantasía preciosa, sensibilidad viva, sabe amar con ternura y con pasión; tiene su alma aptitud maravillosa para todo aquello que del alma depende; pero al mismo tiempo está llena de las supersticiones más groseras; sus ideas religiosas son vagas, monstruosas, equivocadas; sus ideas morales no tienen más guía que el sentido natural" (She possesses beautiful imagination, sharp sensitivity, she knows how to love with tenderness and passion; her soul possesses a wonderful aptitude for everything that is soul related; but at the same time, she is full of vulgar superstition; her religious notions are vague, monstrous, wrong; her moral ideas are guided only by her biological instincts).[22] Golfín indirectly points to the importance of making sacrifices to clear the path toward modernization—positive traits like sensitivity or passion needed to be relinquished in order to privilege the rational force and precise thinking that progress demanded.

Marianela's demise at the end of the novel symbolically encapsulates the central message in Galdós's social foundry: in order to consolidate the basis of the modern nation, national identity needed to be detached from the past. Yet the novel's tone is positive; it shows confidence in the *surgical* capabilities of science and trust in the possibilities of technology to eliminate the impediments to modernization—through invasive intervention, science and technology could positively diagnose and treat the nation's ailments. Hence the extraordinary enthusiasm that Pablo's cure raises—his recovery not only projects hope on the new direction that the country was following but also substantiates the notion of labor as one of modernization's main forces. As a matter of fact, Pablo's inability to actively contribute to society with his work is presented in the novel as a major hindrance: "Para él [para Pablo]—añadió el patriarca de Aldeacorba con profunda tristeza—no existe el goce del trabajo, que es el primero de todos los goces" (For him [Pablo]—Aldecorba's patriarch added with deep sadness—the pleasure of working, which is the most important of all joys, does not exist).[23] It comes as no surprise that throughout the story the narrator describes Pablo as a "vegetable," a "bird with broken wings," or a "rock," thus highlighting "Nature's mistakes" in giving great talents to an individual who is unable to use them for the benefit of society.[24] The negative connotations of Pablo's blindness, moreover, are inextricably linked to Marianela and everything her character conveys. Cured of its ailments (Pablo's disability) and free of attachments to its past (Marianela's burden), the country could use its *new eyes* to confront the challenges of modernization.

Notwithstanding the optimistic tone that scientific success imprints on the novel's plot, Marianela's sacrifice is a deeply mourned tragedy. Admittedly, recovering Pablo's *sight* was not only beneficial, but indispensable for progress. Yet the gains are questionable when measured against the losses—Marianela here emerges as the undisputable protagonist of the story and, as cultural critic Sarah Sierra eloquently condenses, "represents the most disenfranchised victim of industrialization's slow violence."[25] For there is an affinity between the protagonist and nature, it is modernization's destructive force, and not its capacity for reshaping society, that consolidates as the novel's conclusion. In fact, the story's denouement focuses entirely on Marianela's demise and its transformative effects on Socartes, while the narrator provides no information whatsoever on Pablo's or the mining industry's fate after the medical miracle. In this light, Galdós's optimism with regard to the positive impact of progress needs to be carefully considered: he praises science and technology, but at the same time warns his readers about the collateral damages they might produce. Unlike Pablo's, Marianela's *blindness* to progress could not be treated—her death is just another consequence of modernization's inexorable yet coveted incursion in Spain.

One of the unintended consequences of industry's new economies of power and capital accumulation was the emergence of a marked utilitarianism, a situation the novel decries:

> Se ha declamado mucho contra el positivismo de las ciudades, plaga que entre las galas y el esplendor de la cultura, corroe los cimientos morales de la

> sociedad; pero hay una plaga más terrible, y es el positivismo de las aldeas, que petrifica millones de seres, matando en ellos toda ambición noble y encerrándoles en el círculo de una existencia mecánica, brutal y tenebrosa. Hay en nuestras sociedades enemigos muy espantosos, a saber: la especulación, el agio, la metalización del hombre culto, el negocio; pero sobre éstos descuella un monstruo que a la callada destroza más que ninguno: es la codicia del aldeano.[26]
>
> (A lot has been said against the positivism of cities, a plague that, among the pomp and splendor of culture, corrodes the moral foundations of society; but there is an even more terrible plague, and it is the positivism of villages, which petrifies millions of people, killing in them any trace of noble ambition and locking them up in a circle of mechanical, brutal, and sinister existence. There are very horrible enemies in our societies, specifically: speculation, profiteering, monetizing educated people, businesses; but over them, a monster who silently destroys stands out: the villager's greed.)

In this image, the "monsters" of positivism petrify the individual, making him useless for reasoning but ideal for the operation of the industrial apparatus. Indeed, mechanization counteracted immobility, a "petrification" that resulted not from the population's lack of work capacity but from the excessive ambition of a small group of villagers. Galdós's critique was neither a repudiation of industrial development nor a disapproval of mining exploitation. Instead, it was an exhortation to make these activities profitable for society in general so that the sacrifice of the past was not in vain.

In Galdós's social foundry, past and present are contrasted using industrialization's negative and positive effects on society. For that reason, as an idealization, Marianela could be simultaneously unattractive and virtuous, similar to the way in which the mining industry was simultaneously valuable as an economic engine but damaging to nature. Being spaces for the production of meaning, in the novel medicine and engineering not only highlight knowledge and work as two essential forces of modernization but also operate as mediators, even facilitators, in the desecration of tradition. While the surgeon intervenes in the body, the engineer operates on nature. By presenting the body and nature as usable resources, Galdós exposes the individual and collective consequences of progress, also suggesting that the transition between past and present needed to be carried out through technical and scientific labor. Such an ideological position is clearly delineated in the novel's plot, where the conflict between appearance and reality gives shape to the characters and their tragedy. As Jones eloquently summarizes, *Marianela* "shows Pablo Penáguilas leaping into the arms of reality with an enthusiasm so reckless that he does not stop to realize that he is destroying the life of the girl who has previously been everything to him, and for whom he still means everything."[27] This improvidence, however, also delineates the drama of the difficult but necessary awakening of society to the realities of modernization. It is assuming this exigency that in the novel Golfín's initial doubts about Pablo's operation fade.

Optimistic attitudes toward progress, like the one Galdós shows in this text, became more problematic and ambivalent during the following decades, once the plans for social transformation projected by the liberal revolution were defaced to fit the conservative structure of Cánovas's political system. Writers felt the disillusionment of a country trying to restore the past, rather than embracing the future. During the 1880s, and in response to this disenchantment, the thesis novel gave way to a more-pronounced literary realism. Notwithstanding the ideological conflicts posed by this aesthetic turn, literary narrative would evolve into elaborate portraits of society that attempted to capture life itself.[28] In this context, industry continued playing a central role as a space for the codification and interpretation of social conflict. The following section examines how these new literary perspectives implied a revision of Galdós's initial assessment of the country's modernization.

Progress as Dejection in José Ortega Munilla's *El tren directo*

The title of Ortega Munilla's novel *El tren directo, una relación contemporánea* (Express train, a contemporary story) reveals a series of contradictions with regard to modernization that deserve a closer examination. At first glance, one might think that the story deals with the possibilities of the railroad in terms of shortening distances or transforming the perception of the present. Yet the novel's plot does not focus on train travel or its effects on daily life. Rather, it describes the dramatic and tragic events that the engineering and implementation of the railroad trigger in a small village. Far from addressing the operation of the transportation system, as suggested by the title, the novel is a warning about the negative consequences of modernization for rural communities and for the environment. Furthermore, the emphasis placed on the railroad's speed (hence the explicit allusion to this train being "direct" or express) is not translated into the story's narrative structure; on the contrary, the novel seems to deviate constantly from the main plot to expand on the secondary characters' vicissitudes, signaling a metaphor for progress's *multiple stops*. One connecting thread in the argument of *El tren directo*, nonetheless, is the continuous exploration of the low passions—ambition, greed, envy—that economic development promoted in the local population. In this portrait of a morally decomposed society, Ortega Munilla finds the elements to evaluate and criticize the multiple incompatibilities between industrial development and Spanish identity.

By locating the plot in rural Spain, Ortega Munilla also explores the country's complex transition from an agrarian economy to a modern financial system wherein traditional structures of power and class stratification became ineffectual. In revising these tensions, the novel exposes the author's disillusion with the 1868 revolutionary efforts to transform Spain. In this context, technological change functions both as an abstraction, representing the instability of society, and as a concrete disruption that affected the harmony of rural spaces. The novel advances a critique of the country's uneven development and its repercussions on economic inequality, social justice, and nature. This social foundry suggests a needed reappropriation of the past; yet, in contrast to Galdós's novel studied earlier, the sacrifice implied in this process is not perceived as a productive step forward but as the expensive and tragic toll of progress.

Ortega Munilla was an active actor in the political and cultural life of fin-de-siglo Spain. Through both his literary and journalistic production, he exerted a considerable influence on diverse social and political circles. He served, for example, as director of one of the most important newspapers of the time, *El Imparcial* (The independent), a role that allowed him to be in constant contact with political circles. This involvement in state matters would become categorical once he became a member of parliament (*Diputado a Cortes*), a position he held from 1898 to 1910. Despite his association with liberal and progressivist ideas during his youth (close to Krausism and fond of naturalism), by the end of his life Ortega Munilla's ideological postures and political views aligned with a more conservative and traditional understanding of the country's problems. As cultural critic León Bodevin points out, beyond his own disillusionment with the possibilities of a liberal revolution, at some point Ortega became a staunch defender of the royal family, the Church, and the old glories of Spanish history.[29] Part of this discontentedness already appears in his early literary production. In contrast to the Galdós of *Marianela*, Ortega Munilla perceived industrialization as a risky process of social transformation in which the interests of the political and economic elites would prevail over collective needs. *El tren directo*, for example, depicts a country that is not ready to assume the challenges of modernization due in part to the separation (symbolic and physical) that existed between urban and rural spaces.[30] In fact, his solution to the national problems hinged on the coexistence of tradition and progress and the effective interconnection of people.

El tren directo tells the story of María Luisa, a beautiful villager who, after losing her mother at an early age, is forced to marry a wealthy businessman. Before her engagement she had met Genaro, a young man of her age and social status with whom she had fallen in love. This love story, however, did not last long, as Genaro leaves the village in order to become an engineer. Years later, and in the context of the expansion of the national railroad network, Genaro returns to Arijona, his hometown, to supervise the construction of a trainline crossing the town. By then, already a widow and mother of a child with health problems, María Luisa suffers constant intimidation from her in-laws, who have decided to challenge her right to inherit her husband's fortune and to take custody of their daughter, Justina. This conflict becomes even more intense after the lovers' reunion, a situation used as a pretext by Maria Luisa's in-laws to win the legal battle and destroy her life. During the process, the child's health worsens, leading to her death in the middle of María Luisa's attempt to flee injustice. By emphasizing a contentious dynamic of displacement, separation, and reunion in the context of industrial modernization, the novel establishes multiple contrasts between past and present, urban and rural spaces, and reason and emotion, thus showing the impossible convergence of tradition and progress in Spain.

Few recent studies on nineteenth-century cultural production have focused on Ortega Munilla's contributions to literature and politics. The most common line of analysis of his work connects his writings to the conflictive development of realism in Spain, highlighting the author's influence on writers such as Galdós.[31] Other critical approaches have focused on tracing his political affinities—research by Ruth Schmidt, León Bodevin, and Jorge Medina has concluded that his novels, particularly the ones written during the 1880s, present a Romantic view of social conflict that ideologically associates his stories with some of the liberal principles of Krausism.[32] In contrast to those readings, I contend that *El tren directo* is an analysis of the different social structures and power dynamics that emerge in a small village unprepared to embrace technological modernization.

Ortega Munilla's social foundry relies on the novel's narrative structure: the main characters' sentimental conflict, for example, mirrors the social tensions he is evaluating. María Luisa and Genaro's relationship is thus shown as morally harmful—it represents a threat to the stability of traditional social institutions. Paradoxically, Genaro's decision to return to Arijona is not conditioned by his former affection for María Luisa, but by his desire to be with his family while his commission is completed. As engineer in chief, he is allowed to bring his wife and children with him, so he can contribute to the country's material development without sacrificing his personal life or the happiness of his loved ones: "Enriqueta anhelaba también vivir tranquila, y por más que no le gustase gran cosa el meterse en un mísero poblacho, llevaba con agrado el sacrificio, por no verse separada de su marido dos años enteros. Desde que se casaron apenas vivieron juntos seis meses. El ferro-carril, que une a los seres distantes, separaba aquellos dos cuerpos de continuo" (Enriqueta was dreaming of the possibility of living in peace, and even when she did not like the idea of moving to a wretched village, she gladly made this sacrifice so she could be together with her husband for two years. From the date they got married, they had lived in the same place for only six months. The railroad, an invention that brings people together, continuously separated these two souls).[33] Technological advancement, as depicted here, demanded sacrifices that threatened the institution of the family; interestingly, it was engineers' job to find mechanisms to transform society while preserving traditional structures and defending long-established practices.

This important yet contradictory mission seems to be particularly difficult for Ortega's characters. In the novel, modernization is stigmatized as a damaging force: it destabilized the social structure and harmed the natural environment. If rural Spain was a space defined by tradition, then Genaro and María Luisa's reunion represented an affront that not only exposed Spain's unsuitability for accepting industrial transformation but also distorted the engineer's role as an agent of civilization.[34] This tension is also reflected in the relationship between María Luisa and Genaro's wife, Enriqueta—the former Spanish, the latter French; one representing the countryside, past, and tradition; the other a symbol of the city, modernization, and bourgeois society. Associating the rural space with a stereotypical image of backwardness, Enriqueta had formed an image of María Luisa as unsophisticated, unattractive, and ignorant—similar to Marianela's characterization in Galdós's novel:

> Lo único que me quita el miedo es que esa María Luisa será una lugareña basta. . . . Sí, bien segura estoy de ello. Ha de tener una cintura como una tinaja y unas manos gordotas y feas. . . .
>
> Y vestirá muy mal. . . . Y llevará siempre un pañuelo de seda de color chillón en la cabeza. . . . ¿Y el peinado? ¡Dios mío! El peinado será de seguro ese feo rodete que es aquí uso, y que da a las mujeres el aspecto de perros de aguas con las orejas sin esquilar, abrumadas de ricillos de lana.[35]
>
> (The only thing that makes me feel better is that María Luisa is for sure a vulgar villager. . . . Yes, I am sure of that. She has to have a waist of the size of a big jar, and ugly fat hands. . . .
>
> And for sure she dresses badly. . . . And she wears always a garish shawl. . . . And her hairstyle? Oh my God! Her hairstyle is surely that horrible round-pad thing that is so popular around here, and which makes women look like wet dogs whose ears have not been shorn, overwhelmed by little curls.)

The presumed lack of taste and manners of the local is contrasted with the refined urban spirit of the foreigner—in Enriqueta's mind being a villager is synonymous with vulgarity and ignorance. María Luisa, nonetheless, is depicted in the novel as a modern and beautiful woman. In fact, it is this modern character that has gotten her into trouble with her in-laws. Surprisingly, Genaro is not seduced by this misleading image of the past; on the contrary, after his encounter with María Luisa he confirms his intentions to leave the *past behind*, defending the integrity of his family and honoring his position as a champion of progress.

Yet the price of breaking these links is very high: similar to what happens to Marianela in Galdós's novel, the death of María Luisa's daughter, Justina, is presented as the necessary toll of progress. Not coincidentally, the tragic event takes place at the same time as the railroad's opening ceremony. In the description of this moment, the locomotive's whistle mixes with the euphoric yelling of the audience that witnesses the line's inauguration, hiding the pain and sadness of María Luisa's weeping. The triumph of science and technology is celebrated even as the reader becomes aware of some of their harmful consequences, thus exposing two opposite but coexisting sides of progress.

This equivocal nature is also highlighted in the novel to present utilitarianism as one of the risks of misappropriating the advantages of material modernization. Different practices of speculation in the novel distort the economic benefits of progress. Political and economic elites have taken advantage of the legal loopholes of a changing society to keep in place many of the systems of labor exploitation and landowning that prevailed in the feudal economy.[36] Victims of a power struggle over controlling the railroad development and its returns, María Luisa's and her daughter's predicaments are part of these contradictions. The financial interests at stake are contextualized in the story through the monologues of Tío Clavo, Arijona's moneylender and one of the railroad's main investors:

> Hoy he comprado cinco pedazos de tierra, que no valen tres reales cada uno. . . . Pero mañana valdrán tres mil duros como un ochavo. . . . ¡Notable milagro! . . . Para eso no hay más que tirar una línea en un papel blanco, y poner encima: "Esto es un ferro-carril," y este ferro-carril va á pasar por aquí y por allá, y va á cruzar estos cinco pedazos de erial que ha comprado el tío Clavo. . . . "¿Quieren VV., señores ingenieros, que pase por aquí esa mecánica negra que escupe salivazos como un matón, y lleva delante un farol rojo como el fantasma? Pues sírvanse usarcedes pagarme tres mil duros por cada uno de estos pedacitos de arena. ¿Que vale menos? Pues yo no lo doy más barato."[37]
>
> (Today I have bought five plots of land costing less than three reals each. . . . But tomorrow they will cost fifteen thousand pesetas as valuable as any legal coin. . . . What a miracle! . . . To achieve this, you only have to trace a line on a white piece of paper and write on top: "this is a railroad" and this railroad is going to join this and this point, and is going to cross these five plots of land that Tío Clavo just bought. . . . "Do you, dear engineers, want the black machine that blows smoke like a criminal and has a red light in the front like a ghost to cross this land? Well, dear sirs, please pay me fifteen thousand pesetas for each of these grains of sand. Are they worth less? Well, I don't sell them cheaper.")

The railroad is depicted here as an accomplice in a criminal scheme in which a worthless piece of land is turned into a valuable treasure, a cherished possession that feeds greed and justifies any moral deviation of progress.

This particular dislocation of society also illustrates many of the fears and expectations that the country's transformation generated. In this regard, the perspective of Dióscoro, the town's poet (Ortega Munilla's alter ego), becomes relevant for understanding the conflicts that emerge with the railroad's arrival. An analytic, cultivated individual, the poet is able to condense the complexity of these tensions through the use of images and metaphors; his Romantic view thus plays an important part in Ortega's social foundry:

Dióscoro, que no carecía de imaginación, bien que la tuviese percudida y manchada con el contagio romántico, como un lente de aumento que se ha caído en un tintero, comparó allá en sus adentros á aquellos *wagones* con gigantes, y vio en cada uno el símbolo de una preocupación social, de un vicio hereditario de los españoles, de una miseria humana—á todos los cuales llevaba presos y sojuzgados el progreso.—Un *wagon* que, cargado de piedra, al correr daba un ruido sordo, pensó él que era la ignorancia; el otro que seguía luego vacío, más alto que los demás, metiendo mucha bulla de metales removidos y haciendo oscilar sus cadenas como un presumido petimetre que juega con los dijes del reloj, parecióle la fatuidad misma, que heredamos del siglo de los Felipes, sin las grandezas de aquellos buenos conquistadores. Aun presentaba más fisonomía el último carruaje, en donde cien arrobas de carbón de piedra absorbían, sin reflejarla, la luz irresistible del sol primaveral. Vio en él un símbolo del espíritu clerical del país, que cruzaba por la vía echando polvo negro, como quien hecha [*sic*] excomuniones; y por un capricho de la casualidad, los terrones de cok diseñaban la figura de un enorme bonete, debajo del cual, con un trozo de tiza habían pintado unas cifras en el mismo carbón, expresando la cantidad de arrobas que allí marchaban de camino, pareciendo estas cifras la osamenta de un cráneo descomunal y medroso.[38]

(Dióscoro, who had an outstanding imagination, although contaminated with certain romanticism reminiscent of an ink-stained magnifying glass, in his mind compared the train wagons with giants, and saw in each one of them symbols of a social problem, of an inherited vice, of a human disgrace—all of which, subjugated and imprisoned, were being carried by progress. A car loaded with stones, making a hollow sound, was associated with ignorance; the one following it, empty and taller than the rest, making noises of moving metals and playing with its chains like an arrogant dandy plays with his watch, was associated with the vanity inherited from that century when more than one Phillip was on the throne, only without the glory and fortune of those courageous conquerors. Even more notorious was the last car, where twenty-five hundred pounds of coal absorbed, without reflecting it, the irresistible light of the spring. [The poet] saw on it a symbol of clerical spirit, transiting on the railroad while throwing out black dust, like issuing excommunications; by pure chance, the arrangement of coal stones in the wagon resembled a huge biretta, under which, with a piece of chalk, someone has written on the coal itself a cypher indicating the amount of pounds being carried, numbers that looked like a scary and enormous skull.)

Here, the writer evaluates society from a subjective position inspired by technology—"a magnifying glass stained with ink." With precision, Dióscoro enumerates and describes the multiple obstacles to modernization. The locomotive, representing progress, hauls all the weaknesses of national identity: first, ignorance, one of the acutest problems of Spanish society and a main topic in the novel; second, the vanity of a national character fed with empty glories; lastly, the burden of religion, which complicated the advancement of the country by defending the values of national identity through superstition and vacuity.

The novel draws an emotional map of different responses to progress and does so by counterpoising magic with rationality. At least three discernible reactions to industrial and technological transformation can be found in the text: fear of and amazement at the unknown, rejection of progress's negative effects on nature, and a celebration of modernization as a scientific achievement. That is the case, for example, in the lines dedicated to describing a theodolite, a common tool used in engineering: "Una mañana . . . ¡si parece cosa de cuento! . . . en lo más alto del Cerralvo apareció un objeto extraño é inverosímil que los labradores de Arijona de Arriba no supieron nombrar. Era un tubo dorado y coruscante, que puesto sobre tres patas agudas servía para que una porción de señores extranjeros inspeccionara el país con una curiosidad irritante." (One morning . . . like in a fairy tale! . . . on top of the Cerralvo hill a weird and far-fetched object appeared, a device that peasants in high Arijona could not identify. It was a golden and glittered tube that, lying on three pointy legs, was used by a group of foreigners to inspect the land with an irritating curiosity).[39] The use of adjectives such as "glittered," or the description of technicians as "foreigners," sets up the confrontation between reality and imagination. New objects are immediately associated with the supernatural, thus stimulating fear and superstition: "[En Arijona] se creyó a pie juntillas que el anteojo de Cerralvo era el anteojo del diablo, y hubo comadre aspaventera y curiosa que aseguró haber visto salir de aquel tubo dorado rayos y centellas que secaban los sembrados por donde cruzaban, y hundían las casas contra las que asestados iban" (It was assumed that the lenses on the Cerralvo hill were Evil's spectacles, and there was even a loud and curious gossiper who claimed to have seen how a golden and sparky beam was projected out of the tube, withering all the sown fields and sinking all the houses around).[40] Technology's destabilizing capacity is projected here in mystical terms, threatening nature and consequently the delicate social fabric of traditional Spain.

Yet other characters in the novel perceive the same transformation in a more positive way. For them, the railroad was not only an enormous achievement, a proof of human ingenuity, but also the materialization of the joint efforts of society:

> —Pues el boticario dice que no es cosa infernal, sino una invención de los hombres. . . .
>
> —Y el médico añade que eso no es más que un puchero de hierro que se desliza por dos alambres.

> —¡Virgen del Puerto! ¡Vaya un puchero! ¡Pueden cocerse en él habas para una provincia de hambrientos!
>
> El mayoral Requilorio, que había andado mucho por el mundo, daba explicaciones humorísticas del ferro-carril en otro corrillo.
>
> —Es así como muchas casas juntas . . . una calle de casitas chiquirritinas . . . que no tienen más que una cocina que va delante escupiendo humo.[41]
>
> (Well, the pharmacist says that it [the train] is not a diabolical thing but an invention of men. . . .
>
> And the doctor adds that it is nothing else but an iron pot sliding on two wires.
>
> Mother Mary of the Port! What a pot! You can cook beans for a whole starving town in it!
>
> Requilorio, the foreman, who had traveled the world, gave comic explanations to a different group of people:
>
> It is more like many houses joined together . . . a street of small houses . . . with nothing else but a smoky kitchen in front of them.)

Demystifying the fear of the unknown, these informed descriptions aimed to normalize technology as an integral part of daily life. The train is no longer a diabolical invention, or Devil's tool, but only a quotidian artifact like the cooking pot. Associations between the train's operation and concepts of energy and work are thus semantically refashioned so that their complexity is translated into common images and prosaic processes.

The tension between the unknown and the familiar also affects the perception of nature. Here, mythical images resignify industry's environmental impact. The train thus becomes a monster or a demon with the ability to disrupt and destroy the natural world: "Aquella aparición de fuego da hórrido aspecto al paisaje, y al incierto llamear de la locomotora parecen los peñascos frailes inmensos dormidos, y cabezas gigantescas, cuya cabellera de Medusa, hecha de un puñado de lentiscos, se agita pausadamente" (The burning specter gives a horrible aspect to the landscape; the indeterminate flames of the locomotive make the cliffs look like huge sleeping monks whose gigantic heads, made of a handful of flints, slowly agitate their Medusa-like hair).[42] In the contrast between nature and industry, the landscape reflects the distortion of reality—as the locomotive transfers its energy and dynamism to the surroundings, mountains cease to be static.

While many of the descriptions in the novel project a supernatural understanding of technology, there is also an effort to rationalize progress and vindicate scientific knowledge as an indispensable tool for social transformation. This is achieved through the explicit use of physics to explain factors of the industrial economy such as power distribution, labor exploitation, or circulation of capital. In a heated discussion among villagers about the principles of movement behind the train's operation, for example, Arijona's prosecutor synthetizes the technological complexity of the machine into an evocative old Roman maxim: "Los antiguos decían: '*Motus est causa caloris. . . .*' Los modernos decimos: '*Calor est causa motus. . . .*' De donde resulta, por ser el calor lo que mueve y empuja el tren, que la tal invención es añeja. Los antiguos la sabían ya, sólo que la entendían al revés" (Ancient people used to say, '*Motus est causa caloris . . .*' (movement causes heat). Contemporary people say, '*Calor est causa motus . . .*' (heat produces movement), which allows us to conclude that, since heat moves the train, this invention is not new. Ancients already had it, the only thing is that they understood it in the wrong way).[43] Emphasizing the universal character of science, this passage highlights the power of modernization to subvert the past and *correct* how the "ancients" understood nature. Ortega's social foundry points here to an evident dissonance between the past (when nature was misunderstood) and the present (when the moral values that support modernization have been distorted).

In the novel, symptoms of this moral erosion are attributed to the arrival of foreigners to serve as labor in the construction of the railroad line. Associated with a barbarian invasion, workers' particular cultural codes harmed traditional forms of socialization, pushing conservative sectors to condemn industrial development. From that perspective, notions of civilization and barbarism are subverted so that progress, generally understood as a positive expression of rationality, is now seen as a destructive and even illogical force. To confirm this allegorical transposition, the novel links tradition to the sophistication of classic Rome while portraying modernization as an untamed, violent power: "Fué la invasión de los hunnos [*sic*]. La antigua Arijona, la villa vetusta y clásica, á quien los geógrafos contemporáneos de Plinio nombraron *Urbs Gorgonæ* (ciudad de la Gorgona) por una estatua gigantesca labrada en sus muros, y de la que aun se guardan vastos informes, sintióse estremecida en presencia de tal ejército de extranjeros" (It was like the Huns' invasion. The old Arijona, the classic and traditional village, which classic geographers (contemporaries of Pliny) referred to as *Urbs Gorgonæ* (City of Gorgon) because of a gigantic statue carved in its walls and about which many accounts have been carefully archived, was struck by the presence of this foreign army).[44] With irony, the novel shows how modernization could weaken social order—traditional Arijona turns into the *urbe* (center of civilization and example of harmony) that is attacked and dismantled by barbarians:

> El odio al tren, que venía a romper las costumbres de uno de los más reaccionarios rincones de nuestra patria, estrellóse contra los que habían simpatizado con los introductores de la innovación.
>
> —Esa gente desmoralizada—decía un viejo carlista agitando el palo que le ayudaba a llevar por el mundo su gota y sus setenta años—nos trae de fuera el veneno del pecado. Arijona va perdiendo la pureza de sus costumbres.[45]

> (Hatred against the train, whose arrival was damaging the customs of one of the most reactionary corners of our country, clashed with the sympathy some people felt for those who brought such innovation.
>
> Those amoral people—an old Carlist claimed while agitating a stick he used as a cane to support the weight of his seventy years and his arthritis—bring us the poison of sin from outside. Arijona is losing the purity of its customs.)

In contrast to *Marianela*, Ortega Munilla's novel presents a pessimistic portrait of society in which *seeing* the advantages of progress did not necessarily justify a rupture with the past. On the contrary: his social foundry suggests a necessary reappropriation of tradition, an act of cultural and social cleansing that would generate the new values of national identity.

El tren directo's relevance for discussing the country's social and political transition at the end of the century was immediately recognized by important cultural critics of the period. Leopoldo Alas, for instance, highlighted the subtlety with which Ortega Munilla captures the drama of modernization in a simple but well-articulated narrative structure.[46] Although Clarín's commendation was significant, it disregards important elements: beyond its narrative structure, the novel is embedded in a complex relational space (the *material field*) in which industrial modernization plays a central role. The train's impact on society, for example, is not the only aspect of industrialization that is subjected to scrutiny in the story. The text also highlights a more general process of social recomposition caused by the mechanization of artisanal labor. Patricio Güemes, María Luisa's brother-in-law, functions in the novel as testimony to this process. Upon his return from America, Güemes finds a very different town than the one he had left years before:

> Desde antes de llegar al patio se oía el respirar jadeante del vapor y el silbido intermitente de la automóvil. Más cerca, el suelo trepidaba, los cristales temblaban entre sus marcos de plomo, como almas pecadoras en cuerpos enfermizos; las voces humanas se perdían en el confuso rumor de la alborotante maquinaria. Parecía que mil patas de hierro acoceaban furiosas, que cien caballos salvajes corrían relinchando, que la descomunal bestia de las labores fatigosas se revolcaba ebria y convulsa bajo los cimientos del edificio. Más cerca . . . ya se notaba el calor de la lumbre en el fondo de las cuevas. Desde cualquiera de las ventanas veíanse ojos siniestros e inflamados—que no otra cosa parecían las bocas de los hornos—é iluminados por su foco lumínico, brazos nervudos y vellosos, desnudos, recios, llenos de un titánico vigor, que se adivinaba en los tendones tirantes bajo la piel, como el armazón metálico bajo la tela del maniquí se adivina.[47]

> (Before arriving in the courtyard, one could hear the heavy breathing of the steam and the intermittent whistle of the engine. Closer, one could feel the

> floor trembling, the windowpanes vibrating inside their lead frames, like sinners' souls in sickly bodies; human voices faded in the murmur of the noisy machinery. It looked as if one thousand iron legs were violently kicking, as if one hundred horses were running and neighing, as if the hard-working colossal beast was rolling around, drunk and convulsing under the building's foundations. Closer . . . one could notice the furnace's heat coming from the bottom of the caves. Through the windows, one could see sinister and swollen eyes—that was how the entrance of the ovens looked—and enlightened by the shining spotlight there were hairy, sinewy, naked, strong arms filled with a titanic vigor that was visible in the tight tendons under the skin, like the metallic frame is visible under a mannequin's clothing.)

The narrator turns his attention to the interaction between workers and industry: laborers and objects are homogenized by the machine's operation, and mechanical devices are associated with animals or depicted as sick bodies—commonplaces in literary representations of the industrial dynamism. In spiritual terms, the factory has a "sinner's" soul whose immoral acts are expiated through the disease. In the sensorial aspect, the machinery's noise makes people's labor indistinguishable from mechanical activity. Finally, from a social perspective, images of the energy source that supports this *beast* resemble the vigilant eyes of a monster that not only supervises the workers' labor, but also feeds on it.

Modernization's elevated human and environmental cost, as discussed throughout this chapter, constitutes a recurrent trope in the literature of the period—progress is expensive and presupposes multiple sacrifices, an idea explicitly developed in Tío Clavo's reflections on the economic viability of the railroad: "¡Ustedes quieren progresar! . . . Bien, me parece cosa excelente, pero . . . ¡anden VV! . . . paguen el progreso y así progresaremos todos" (So, you all say you want progress! . . . Well, I think that is excellent, but first go ahead . . . and pay the price of progress, and in that way we all will progress).[48] By dehumanizing labor, industry's complexity is reduced to a profitable material enterprise. Workers, once again, are seen as disposable cogs in the larger machine of capital production.

Although Ortega Munilla's novel does not focus specifically on labor problems, its reflections on the high cost of progress confirm the role of workers in a society that resisted social change but at the same time profited from it. In many respects, his social foundry reveals a contradictory tension between optimism for and fear of industrial transformation. The implementation of the railroad in the novel seems to dissipate this hesitation by pushing society to adapt to the new conditions of modernity. However, by depicting a dystopian reality in which rationality fails to prevent the hegemonic consolidation of the aristocracy and the Church within these new political and economic spaces, Ortega Munilla's novel questions the entire project of national modernization. In fact, the emergence of large industrial centers in rural areas would produce new ways of seeing and understanding such an undertaking. Environmental destruction generated nostalgic views of the past wherein the natural world embodied an idealized image of the nation. The last section of this chapter explores the conservative perspective of Armando Palacio Valdés's novel *La aldea perdida* and analyzes its portrait of industrial development as a profanation of tradition.

Industrial Threat and Lost Paradises in Armando Palacio Valdés's *La aldea perdida*

As other authors studied here, Armando Palacio Valdés was a prolific writer, a dedicated journalist, and a sharp critic of his time. As an artist devoted to portraying the transformation of society, most of his writings focus on the encounter of modernity and tradition in rural Spain. Set in his native Asturias, his work addresses topics such as social-class tensions, economic transition, moral and ethical decomposition, labor exploitation, social mobility, and, of course, environmental destruction. Aware of progress's intangible implications, in *La aldea perdida* (1903) the author reflects on these problems by juxtaposing an archaic and idealized vision of the past with an apocalyptic image of the present. As a product of its time, the novel is imbued by a modernist spirit, which, as ecocritic Kelly Sultzbach notes, informs "a desire to reclaim a sensitivity to heritage and local place."[49] In doing so, it also transcends the idea of "environment-as-object" to acknowledge the notion of "environment-as-being." Following a narrative structure based on the form of Greek epics, the story contrasts an Arcadian model of rural labor with the industrial dystopia of the mining economy. Palacio Valdés's proposal thereby consists of revisiting the past before confronting the future—in contrast to other texts studied in this chapter, in which authors see modernization with certain optimism and try to understand industry's social and material transformations, the novel rejects the present in defense of a nostalgic view of Spain.

La aldea perdida was not the first novel in which Palacio Valdés incorporated the mining world. Years before completing what he considered his best work,[50] the Asturian writer had touched on the problem of labor exploitation in *La espuma* (*Froth*) (1890). Confronting the moral decline of a society infatuated with social status, money, and power, the novel exposes many of the vices that in his view were preventing economic stabilization and social progress: double morals, corruption, and greed, among others, characterize the different actors in this naturalistic representation of fin-de-siglo Spain. As cultural critic Brian Dendle explains, *La espuma* "vigorously satirizes the vulgarity and immorality of the upper classes; the novel is anticlerical in its hostility to the religious practices of Madrid society and reveals considerable class feelings in its rendering of the suffering of miners."[51] To be sure, the novel questions the capacity of the bourgeoisie to restrain the aristocracy's immorality and the Church's ambition, as well as its inability to promote the necessary reforms for potentializing the country's modernization.

An episode in the novel that perfectly illustrates these tensions takes place during the visit of the main character, the affluent Duke of Requena, his family, and close friends to one of the aristocrat's mercury mines. This excursion becomes a pretext for the narrator to offer diverse interpretations of the mining boom and its social and health consequences for miners:

> A la puerta del parque y en las inmediaciones había una muchedumbre que saludó á la comitiva con vivas apagados. Eran los obreros. . . . Todos ellos tenían la tez pálida, terrosa, los ojos mortecinos, y en sus movimientos podía observarse, aun sin aproximarse mucho, cierta indecisión que de cerca se convertía en temblor. La brillante comitiva llegó á tocar aquella legión de fantasmas (porque tales parecían a la luz moribunda de la tarde): los ojos de las hermosas y de los elegantes se encontraron con los de los mineros, y si hemos de ser verídicos diremos que de aquel choque no brotó una chispa de simpatía. Detrás de la sonrisa forzada y triste de los trabajadores, un hombre observador podía leer bien claro la hostilidad.[52]
>
> (Around the depot's door there was a crowd welcoming, with spiritless cheers, the visitors' procession. It was the miners. . . . All of them had pale, earthy faces, dying eyes, and in their movements it was possible to see, even from a distance, some kind of indecision. With a closer look, this indecision was in reality trembling. The illustrious group of visitors made contact with that legion of ghosts (because that was how they looked in the afternoon's dying light): the eyes of the elegant and beautiful visitors encountered the miners' eyes, and if one has to be honest, it could be said that the encounter did not produce a spark of sympathy. Under the fake and sad smiles of workers, an attentive person could have clearly read hostility.)

The visitors' reactions to this encounter varied. The duke, for example, saw in the sickly look of the miners an indication of their moral deviance and condemnable nature:

> Lo que hace aquí falta ¡pero mucha falta! es moralidad. Moralicen ustedes al obrero y todos estos estragos que ustedes han visto desaparecerán. Que no beban, que no jueguen, que no malgasten el jornal, y esos efectos del mercurio no serán para ellos funestos. . . . Pero, claro está,—añadió volviéndose hacia los caballeros que se habían acercado:—¿cómo ha de resistir en la mina un cuerpo que en vez de alimento, sea el que sea, tiene dentro un jarro de aguardiente amílico? Estoy convencido de que la mayor parte de las enfermedades que aquí hay son borracheras crónicas. Sepan ustedes, señores, que en Riosa se desconoce por completo el ahorro, . . . sin el cual no es posible el bienestar ni la prosperidad de un país.[53]

(What we need here, but with urgency! is morality. Teach morals to the worker and all the ravages that you have seen will disappear. May they stop drinking, stop gambling, stop wasting the salary, and mercury effects would not be lethal for them. . . . But, of course, he added turning toward the gentlemen who had approached: how would a body that instead of food has alcohol inside resist the work in the mine? I am convinced that most of local diseases are just the result of chronic drunkenness. Know, dear sirs, that in Riosa saving money is completely unknown, . . . and without it society's well-being and prosperity is impossible.)

Requena's utilitarian vision exemplifies one of the many ways in which the aristocracy appropriated progress and manipulated it for their own benefit. Spanish elites, as historian and cultural critic Guillermo Gortázar explains, never let ideology interfere in their financial and corporate undertakings—it was not uncommon to find recognized advocates of conservatism participating in modern industrial and financial enterprises as partners of well-known liberals or republicans.[54] For a writer like Palacio Valdés, the materialistic nature of this contradictory behavior was reprovable—*La espuma*, in fact, presents an elaborate critique of such social dysfunctionality.

In spite of Palacio Valdés's commitment to exposing the vices of society in his novel, the story does not give workers any agency or centrality. On the contrary, it is the upper classes and the bourgeoisie that bear the plot's dramatic weight. The figure of the medical practitioner, for example, is central to ideas of diagnosing and prescribing the ailing nation. The comparison he makes in the novel between the behavioral pattern of microorganisms in a drop of water and the different actors in society attests to this particular characterization:

Quiroga (que así se llamaba el médico) concluyó mostrándoles una gota de agua. Uno por uno todos fueron contemplando el mundo invisible que dentro de ella existe.

—Veo un animal mayor que los otros, —manifestó el duque, aplicando con afán uno de sus grandes ojos saltones al agujerito del aparato.

—Observará V. que delante de él todos los demás huyen, —dijo el médico.

—Es cierto.

—Ese animal se llama el rotífero. Es el tiburón de la gota de agua. . . .

—Es la historia de siempre. En la gota de agua, como en el mar, como en todas partes, el pez grande se traga al chico.[55]

(Quiroga, such was the name of the doctor, finished his demonstration by showing them a drop of water. One by one, the visitors contemplated the invisible world contained in it.

I can see an animal bigger than the others, the duke utters, eagerly pressing one of his protruding eyes against the small hole of the instrument.

Do you see how in its presence all the other creatures run away? the doctor says.

That is right!

That is a rotifer. It is the shark of the drop of water. . . .

It is the same old story. In the drop of water, like in the ocean, like everywhere else, the big fish devours the small one.)

While the doctor understands social inequality as problematic, he accepts it as part of a natural process. Power unbalance, once again (see analyses in previous chapters), is presented as a natural and necessary condition for the functioning of society.

Negative depictions of both the aristocracy and the proletariat are symptomatic of the author's profound distrust of the capacity of industrial modernization to generate positive outcomes. Yet the novel's skepticism toward social transformation is quite moderate in contrast with the pessimism exhibited in later works. Palacio Valdés's decision to return to the practice of dogmatic Catholicism after his second nuptials in 1899 would be in part responsible for this change in his approach to national progress.[56] Shifting from analysis and criticism of the present to a nostalgic view of the past, in *La aldea perdida*'s social foundry national regeneration is subjected to the recovery of moral and religious values. By focusing on the negative effects that the arrival of miners generated in the countryside, the work denounces the desecration of tradition, a process that has both physical and symbolic consequences. Formerly represented as victims of the misappropriation of progress by the social elite, miners now become instruments of this deviation, and as such, personifications of the most perverse aspects of industrial expansion. In order to emphasize this negative image, the novel re-creates a rural Arcadia whose harmonious way of living is dismantled by the emergence of new economies of production.[57]

The society depicted in *La aldea perdida*, therefore, hardly corresponds with reality. On the one hand, the classic epic's evocation situates the village out of time: elements like heroism, honor, or loyalty constitute the pillars of this imagined community. One of the jobs of the heroes in this story is to gain the love of the female villagers to later form the families that will maintain the prosperity and perdurability of the region. As part of this idealization, the novel even attempts to capture the oral tone of epic poetry, chanting the deeds and adventures of its protagonists: "¡Y aún no ha cantado [el poeta] á los héroes de mi infancia! ¡Aun no te ha cantado, magnánimo Nolo! ¡Ni á ti, intrépido Celso! ¡Ni á ti, ingenioso Quino! ¡Aún no ha caído a tus pies, bella Demetria, la flor más espléndida que brotó de los campos de mi tierra!" (And the Poet has not yet sung to the heroes of my childhood! He has not yet sung about you, magnanimous Nolo! Nor about you, fearless Celso! Neither about you, ingenious Quino! Not yet has he fallen on your feet, beautiful Demetria, the most splendid flower blooming in the fields of my land!)[58] On the other hand, the novel pinpoints the environmental and moral effects of industrial mining as the sole cause of the destruction of this atemporal paradise, ignoring other important dimensions of the problem.

In condemning the disarray of the countryside as the result of the miners' arrival, nonetheless, Palacio Valdés takes a clear political stand on the labor question. In contrast to other authors studied in this book, his social foundry entails a very conservative, traditionalist codification of industrial change. Images and metaphors that in other representations of progress were used to contrast past and present—light and darkness, good and evil, or purity and contamination—now denote a mystical space disrupted by historical and material forces. The color of coal, for instance, is employed as a marker to symbolize the antiheroic character of the story's two antagonists: miners Plutón and Joyana. Contrary to the villagers, they embody social vices such as cowardice, disloyalty, and lust: "Pero lo más negro de todo lo negro que había en Laviana era Plutón. Aquel hombre ya no era hombre, sino un pedazo de carbón con brazos y piernas. . . . Lo mismo le importaba a aquel malvado dar una puñalada que beberse una copa de aguardiente" (But the darkest of all the dark things in Laviana was Pluto. He was not a person anymore, but just a piece of coal with arms and legs. . . . For him it was the same stabbing someone or drinking a shot of aguardiente).[59] In fact, it is Plutón's condition as a dark inanimate object associated with industry ("piece of coal") that explains his moral deviance—he personifies the decline of a society in which industrial workers were at the same time victims of modernization and victimizers of the past.

In her analysis of Palacio Valdés's work, cultural critic Guadalupe Gómez Ferrer sees this resignification of the miner—from an anonymous character (*La espuma*) to a stigmatized criminal—as a response not only to the increasing notoriety of the labor conflict, but also to the author's leaning toward conservatism.[60] It is therefore significant that in *La aldea perdida* miners are identified by their names or epithets, and that those names have allegorical connotations. In introducing Plutón, for example, the narrator adds: "Mr. Jacobi, ingeniero alemán, director de la explotación, hombre letrado y no poco bromista, comenzó á llamarle Plutón por haber nacido debajo de la tierra, y Plutón se quedó" (Mr. Jacobi, German engineer and director of the mine, an educated man prone to joking, started calling him Pluto because he was born underground, and from that moment Pluto became his name).[61] Mythological associations in the novel also operate as moralizing forces—in classic literature the story of Pluto is connected to the tension between reason and instinct. In Roman theogony, for example, Pluto is the god of the underworld, the place where human transgressions are judged. These symbolic correlations highlight the *descent* of society toward a destructive space marked by the economic dynamism of industry.

Palacio Valdés opposes the harmonious nature of the past to the instability of the present to offer a wide range of perspectives on modernization—from promoters of progress to defenders of tradition. Characters such as Antero Ramírez, "promovedor incansable de los intereses de aquella región y apóstol elocuente del progreso" (indefatigable guardian of the region's interests and eloquent disciple of progress),[62] for example, propose their cases in defense of Laviana's future:

> Amaneció al cabo el día por nosotros tan ansiado, el día en que nuestro valle salga de su profundo y secular letargo. Aquellos tesoros que nuestros padres pisaron siglos y siglos sin sospechar su existencia, para nosotros los amontonó la naturaleza debajo del suelo. . . . Dentro de pocos meses oiréis resonar por estas montañas el agudo silbido de la locomotora. Es la voz del vapor que nos llama á la civilización.[63]
>
> (The day everybody was waiting for finally arrived, the day in which our valley will leave its profound and secular lethargy behind. Nature accumulated for us, under the soil, those treasures our fathers walked on for centuries and centuries without noticing their existence. . . . In a few months, you will hear the locomotive's whistle reverberating through these mountains. It is the steam's voice calling us to join civilization.)

From this perspective, the call of progress could no longer be ignored. Obtaining the land's real wealth did not depend anymore on traditional agriculture; it would require the development of industrial infrastructure. In any case, none of that would be possible without a collective effort: "Para llegar a tal grado de civilización era necesario que los lavianeses aunaran sus esfuerzos" (To achieve such a level of civilization it was necessary that all people from Laviana combined their efforts).[64] Society, as a consequence, could only advance if there was a common will to change.

Yet achieving a consensus of opinion regarding modernization seems to be impossible in Laviana. For many villagers the "renovation" of the country implied a figurative rupture with tradition, and therefore an unwelcome change in their way of living: "Vagamente todos sentían que una transformación inmensa, completa, se iba a operar pronto en Laviana. El mundo antiguo, un mundo apacible y patriarcal que había durado miles de años, iba a terminar, y otro mundo, un mundo nuevo, ruidoso, industrial y traficante, se posesionaría de aquellas verdes praderas y de aquellas altas montañas" (Vaguely, everybody perceived that a complete, immense transformation was soon going to take place in Laviana. The old world, a peaceful and patriarchal world that had lasted for thousands of years, was coming to an end, and a different world, a new, noisy, industrial, and commercial world would take over those high mountains' green prairies).[65] In the novel negative visions of modernization prevail over those that highlight its benefits. In general, Palacio Valdés's social foundry shows modernization as a threat to traditional, high moral values. Echoing this opinion, Laviana's patriarch, Don Félix Cantalicio Ramírez del Valle, warns his fellow countrymen about the high costs of progress:

> Hasta ahora hemos vivido a gusto en este valle sin minas, sin humo de chimeneas ni estruendo de maquinaria. La vega nos ha dado maíz suficiente para comer borona todo el año, bien sabrosas patatas y legumbres. . . . El ganado nos da leche y manteca, y carne si la necesitamos, tenemos castañas abundantes que alimentan más que la borona y nos la ahorran durante muchos días, y esos avellanos que crecen en los setos de nuestros prados producen una fruta que vendida a los ingleses hace caer en nuestros bolsillos todos los años algunos doblones de oro. ¿Para qué buscar debajo de la tierra lo que encima de ella nos concede la Providencia: alimento, vestido, aire puro, luz y leña para comer nuestro pote y calentarnos en los días de riguroso invierno?[66]

> (Until now, we have pleasantly lived in this valley without the mines, without the chimneys' smoke, or the machinery's noise. The land has given us enough corn to eat bread all year long, very tasty potatoes and greens. . . . Cattle provide us with milk and butter, and meat if we need it; we have plenty of chestnuts, which provide more sustenance than bread and keep us from just having bread for days. And those hazelnut trees growing on the fences of our lawns produce a fruit that, when sold to the British, put in our pockets some gold coins every year. What is the purpose of looking under the ground what the Providence already placed for us on top: food, clothes, pure air, light, and wood to cook our food and keep us warm in winter?)

Don Félix's contrast between the surface and the underground functions as an allegory of the tension between past and present, tradition and progress. However, the vision of a luminous, bright countryside that prevails over the darkness of industrialization in this image is only possible in a static, isolated, and atemporal Laviana.

In this hypothetical isolation, intrusions from the external world are devastating. The construction of industrial complexes and transportation networks in the village, for example, has grave repercussions not only for the dynamics of social organization but also for the order of nature:

> Mas los prados, los árboles y los seres vivientes que se agitaban en aquel delicioso paisaje no recibían con igual satisfacción la visita del huésped. Sus penetrantes silbos estremecían la campiña. Volaban los pájaros, corrían las reses hasta despeñarse, huían los niños, ladraban los perros en los caseríos, ¡como si en vez del bienestar y la riqueza les trajese aquel glorioso artefacto la oscuridad, la maldición y la guerra! Y los conspicuos, al ver la general desbandada, reían llenos de lástima y excitaban al maquinista para que hiciese

> más ruido, gozándose como los antiguos conquistadores con el espanto que su paso producía.[67]
>
> (Yet grasslands, trees, and creatures that moved around the rich landscape were not pleased with the visitor's arrival. Its penetrating whistle disturbed the prairie: birds flew away, cattle ran until they fell off the cliff, children ran away; in the plantations, dogs barked as if instead of wellness and wealth, that glorious artifact were bringing darkness, curses, and war! To complete this picture, some perverse passengers, when seeing this general stampede, laughed with commiseration and instigated the machinist to make even more noise, enjoying, like old conquerors, the fear that their presence produced.)

Industrialization is described here through warlike images in which negative and dark forces suddenly invade and *conquer* natural spaces. Similar to depictions of technology in Ortega Munilla's novel discussed earlier, the allusion to "old conquerors" produces a symbolic inversion in the classic confrontation between civilization and barbarism. Nature, miners' immorality, and mining itself are all treated as characters, thus emphasizing a sort of ecological consciousness that also has implications for discussions on the relationship between humans and the larger environment.

In this context, the novel turns again to contrasts between light and darkness to present mining casting a shadow over the rural Arcadia. Once more it is Don Félix, lieutenant colonel of the Royal Guard, defender of tradition and, as the narrator emphasizes, staunch admirer of the country's past, who warns his fellow countrymen about this negative aspect of progress: "Los hombres trabajarán más que antes y no á luz del día y respirando la gracia de Dios como ahora, sino metidos en negros, inmundos agujeros. Las mujeres lavarán más ropa sucia, cuidarán más enfermos, quedarán viudas primero. Los niños escucharán más blasfemias, sufrirán más golpes. Yo me río de esa prosperidad y la maldigo" (Men will work more than before, and not under the sunlight or breathing God's grace, like they do now, but inside dark, filthy holes. Women will wash more dirty clothes, take care of more sick people, become widows earlier. Children will hear more swear words, be punished more often. I laugh at that prosperity and curse it).[68] As a counterproductive enterprise, industrialization is associated here with a contaminating force that not only debases the environment, but also spoils the joy of labor.

For Palacio Valdés, in rural areas industrial modernization had facilitated the normalization of damaging conducts, fostering, among other negative attitudes, sentiments of greed and ambition. In the novel, this social disarray mirrors the environmental impact of mining activity: "El valle de Laviana se trasformaba. Bocas de minas que fluían la codiciada hulla manchando de negro los prados vecinos; alambres, terraplenes, vagonetas, lavaderos; el río corriendo agua sucia; los castañares talados; fraguas que vomitaban mucho humo espeso esperando que pronto las sustituirían grandes fábricas que vomitarían humo más espeso todavía" (The Laviana Valley was changing. There were mine caves from which the coveted coal flowed, staining with black the surrounding lawns; wires, embankments, railway trucks, sinks; dirty water running through the river; chestnut trees cut down; forges vomiting dark, thick smoke, that soon would be replaced by big factories that would vomit even thicker smoke).[69] Reappearing here as a critique of the materialistic longing for wealth, the image of a *polluted* society projects modernization as an ailment that affected both nature and the moral dimension of identity.

Given *La aldea perdida*'s resemblance to a classic tragedy, Palacio Valdés's social foundry also entails an evaluation of ethical principles. Pluto's myth, for example, functions not only as a frame of reference to allegorize the miners' deviant character but also as a marker to measure the country's moral decline. Seeing industrial development as a desecration of tradition, the pagan image of the underworld's deity highlights the problems of centering the project of modernization on material growth. In his defense of society's traditional spirit, the town's poet, César de las Matas de Arbín, expands on the consequences of this approach:

> Demetria ha muerto y se prepara el advenimiento de un nuevo reinado, el reinado de Plutón. Saludémosle con respeto, ya que no con amor. . . . ¡Con amor no! Yo no puedo amar á ese dios subterráneo que ennegrece los rostros y no pocas veces también las conciencias. La Arcadia ha concluido. Esta raza sencilla y belicosa de nuestros campos desaparecerá en breve y será sustituida por otra criada en el amor de las riquezas y en el orgullo. . . . Desde sus viviendas suntuosas unos hombres de la nada, hijos de labriegos y menestrales, me señalan con el dedo a sus vecinos haciendo escarnio de mi figura y mi pobreza. . . . La lucha es imposible amado primo. A la aristocracia sucede la plutocracia.[70]
>
> (Demeter has died, and a new kingdom is coming, the kingdom of Pluto. Let us greet him with respect, but not with love. . . . Never with love! I cannot love this god of the underground that darkens faces and, not on a few occasions, also consciousness. Arcadia is over. The unsophisticated and belligerent race of our prairies will soon disappear and be substituted for a new one raised in the love of wealth and pride. . . . From their sumptuous houses, simple men, sons of peasants and farmers, will draw attention to their neighbors by pointing at me, making fun of my looks and my poverty. . . . The battle is lost . . . aristocracy will be replaced by plutocracy.)

This description situates Demeter, both the mythological figure and the character in the novel, in the middle of two irreconcilable realities—rural Arcadia and industrial society. Demeter, goddess of agriculture and representation of nature, is sacrificed to the accumulation of capital, a concrete expression of that which

Pluto symbolizes and one of the economic substructures that supported industry. It comes as no surprise, then, that at the end of the novel Pluto, the miner, literally kills Demeter, thus materializing Don César's allegory and confirming the actual horrors of modernization.[71]

As I have shown in previous analyses, industry's dynamism subverted rural immobility. However, in the novel greed derails the potential benefits of this transformative capacity by dissociating national modernization from social development. This disconnection also appears on the narrative's inconsistencies between form and content, namely between the plot's classic note and the actual representations of mining and its different actors. Once again, it is Don César who calls attention to the risks involved in misunderstanding progress:

> Ignoro, señores míos todos y muy queridos amigos algunos, si esos que llamáis progresos industriales van tan estrechamente unidos a la causa de la civilización como os complacéis en suponer. . . . Quiero decir que lo que se nos ha dado como un medio lo convertimos en fin. De aquí se origina siempre un grave desequilibrio, que engendra la corrupción y los vicios. . . . Temo en conciencia ¡oh señores míos! que confundáis lamentablemente la civilización con el industrialismo. Yo sé de países muy industriales donde la cultura del espíritu no corre parejas [*sic*] con las comodidades y refinamientos de la vida. Penetrad en una de las ciudades fabriles de Francia ó de Inglaterra. ¡Cuán suntuosas son aquellas viviendas! ¡cuán delicados los manjares que allí se gustan! ¡cuán blandos los lechos! ¡cuánto pormenor delicado que halaga la vida corporal! . . . Pero escuchad á aquellos hombres en sus refectorios, en sus cafés y en sus teatros, y tengo por seguro que no quedaréis maravillados ni de la agudeza de su ingenio, ni de la elevación de su espíritu.[72]

> (I do not know, dear friends, if those that you define as industrial advances are really linked to the central goal of civilization, as you seem to be pleased in presuming. . . . What I mean is that we have been blessed with a means to achieve things, but we have turned it into an end. This way of seeing things generates a disbalance, creating corruption and vices. . . . I am very afraid, my dear friends, that you are sadly confusing civilization and industrialization. I know of many industrial countries in which spiritual improvements are not in line with the level of comfort or other improvements in people's lifestyles. Visit an industrial city in France or England: see how sumptuous those houses are! How delicate the food they eat is! How soft their beds are! How many details adorn corporeal life! . . . But pay attention to what those men have to say when they gather in restaurants, coffee houses, or theaters, and I am sure you will not be amazed with their wit or the elevation of their spirit.)

In Don César's view, the material comfort industrialization may offer is empty of spirituality and culture. If modernization were to bring great economic and social advantages, the cost of those improvements was too high for the preservation of tradition.

As Palacio Valdés makes clear through the reflections of his characters, in rural Spain industrialization was not the engine of civilization; on the contrary, it appeared to be an agent of barbarism. The ending of *La aldea perdida* synthesizes this idea by showing the tragic consequences of sacrificing the past in the name of progress. In contrast to what Galdós proposes in *Marianela*, and more in line with what Ortega Munilla suggests in *El tren directo*, here violence is the only way in which the new logic of progress can be successfully imposed on the country. In fact, the assassination of the main characters on their wedding day destroys any possibility of preserving and multiplying, through the formation of a family, the moral and religious principles of traditional Spain.

Conclusion

THE UNREACHABLE FUTURE

Spain's industrial boom at the turn of the twentieth century attests to the country's singular process of modernization. With evident allegorical consequences, all-pervasive industry's technological, scientific, and operational models also altered how people interpreted and defined reality. In the field of discursive production, this appropriation of industrial modernization as a cognitive and textual referent was intrinsically connected to the country's historical conflicts. Writing thus constituted a privileged space for exploring industry's impact on the complex negotiation of the past amid the construction of a modern national identity. In a society that defined its ethos in relation to tradition, industry's transformative capacity was extremely problematic. The most important indicators of progress, such as education, innovation, and social mobility, did not align with traditional religious and social principles that were deemed the essence of national identity. As a consequence, industrialization was perceived as a threat—the emergence and consolidation of the working class, the strengthening of peripheral nationalisms in industrial areas, and the 1898 military disaster further complicated this perception. At the same time, this scientific and technological revolution had profound effects on people's mechanisms of socialization. All these elements contributed to the clash of temporalities (past and present, tradition and progress) that shaped the various *social foundries* discussed in this book.

Spain's struggle to synthesize its projects of industrial modernization with a national identity conceived as part of a sacred, spiritual, and static tradition yielded the perception that the future was elusive, unreachable, or ill-conceived. Levels of confidence in the possibility of national transformation varied in accordance with particular historical circumstances. In the euphoria that followed the 1868 liberal revolution, for example, political optimism led to the belief that Spanish society was ready to break with its past, and embrace, even celebrate, social change. Texts written around this period appropriated the language of science and industry to imagine a society wherein new notions of family, class stratification, and labor needed to be defended at the cost of sacrificing the past and tradition. This optimism was also reflected in different proposals for creating a system of higher education that promoted scientific innovation and technological development while harmonizing the practical and theoretical aspects of academic instruction. Society actively engaged with these changes, and the people's joint efforts were thought to be essential to the country's push toward modernity. Yet this optimism did not last long. The emergence of a collective sentiment against modernization, in which the social transformation of industry was seen as a threat to traditional values, revealed the country's political instability and economic crisis, a situation that worsened after the 1898 war with the United States. Idealizing the past, social foundries produced under this pessimistic lens presented progress as an evil and destructive force. Tradition, as a consequence, needed to be restored in order to facilitate national regeneration. In these readings of the country's situation, the future was elusive, confirming once again the contradictory nature of modernization—it seemed as though in order to advance into the future it was necessary to look back to the past.

But refashioning tradition in the context of modernization was not the only way of approaching the question of national identity. Dealing with industry's social and economic transformations also implied a cognitive negotiation of the notion of progress. Authors' preoccupation with how industrialization was assimilated was displaced by a concern with the causes of its misappropriation, namely the deviation from the collective benefits it could produce to favor the interest of certain social actors. This perspective brought into question society's ability to embrace modernization insofar as national identity continued to be ascribed to old models of ideological oppression and class exploitation. Industrialization, as the material realization of these ideals, was thus seen as a modern instrument of class, religious, and political subjugation. According to this reading, technological and scientific development, as well as industrial control, had been co-opted by the elites, who, in addition to utilizing social difference to promote nationalistic animadversion, in many cases also tolerated negative traits, such as hypocrisy, greed, or disloyalty, as natural expressions of the materialistic spirit of the times. Social foundries built on this disposition tend to be ambiguous in their criticism of modernization, which they often present as a necessary evil.

This multiplicity of discourses illustrate the plasticity of industrialization as a metaphorical referent. Scientific popularization, political essay, journalism, and literary fiction are among the many written expressions that confirm textual production as an ideal space for negotiating the complexities of modernization.

The assessments of society proposed in the various texts analyzed throughout this book attest to the authors' positions as intellectuals within a specific cognitive, social, political, and economic context—what I have termed *material field*. In this regard, their work constitutes a testimony to the multiple, conflictive emotions that industrial modernization produced in *fin-de-siglo* Spain. It also shows to what extent writers, politicians, educators, and scientists were involved in the consolidation of the ideological and material conditions that sustained national development.

Among the multiple changes these thinkers desired for the modern country, refashioning the system of higher education stood out as a priority. A lack of political will and limitations of intellectual and material capacities, however, made the proposed educational models practically ineffectual for reconceiving and promoting scientific discovery and technological implementation. The consolidation of a national science was thus obstructed, to a large extent, by the state's reluctance to support a secular vision of society that endorsed industrialization's economic and social redistribution of capital. Through their dedication and tireless efforts, however, educators and scientists aimed to shape collective perceptions of industrial transformation not as a threat to identity but as an engine of national modernization. Fully aware of the importance of leaving the past behind to embrace the dynamism of the present, these intellectuals were sometimes ahead of their time in projecting a view of a country in which science, technology, and education were the main agents of progress. Their work therefore led to the consolidation of a scientific work ethic and to the palpable improvement of the educational apparatus. Furthermore, in most cases the individual achievements of these scientists set a path for their disciples that continues to the present. Even though the fruits of these educators' commitment would not be tangible until later in the twentieth century, their work revealed the capital importance of science, technology, and industry in the country's modernization.

As a matter of fact, the process of cultural mythification of those three elements generated the cognitive conditions that led to the emergence and proliferation of social diagnoses like the ones encapsulated in the social foundry. To a great extent, the way in which this appropriation took place responded to society's tendency to naturalize the unknown. Not surprisingly, this mechanism of cultural normalization was conditioned by the complexities of the social context within which it operated. In the case of Spain, religion was one of the defining factors in the assimilation of industrial modernization. In one way or another, all the texts studied here propose strategies for reconciling the materialistic aspect of science with the spiritual character of national identity. Krausism was instrumental in this attempt to provide social foundries with a coherent narrative, largely framed within three different approaches. For some authors, rationality in Spain was tainted by a proclivity to imagination, individualism, and passion, a tendency that delayed collective efforts to materialize progress. Other thinkers believed that the inability to produce original knowledge and consolidate the bases for scientific and technological development resulted from society's attachment to traditional political structures, economic models, and social organizations. Paradoxically, the third group blamed the backwardness of the country on the exact opposite: the nation's continuing distancing from its past. Regardless of their approach, none of the authors studied here doubted that science, education, and technology were key to industrial development and national modernization, nor did they ignore the need to recover, preserve, or refashion certain aspects of tradition to modernize national identity. It was precisely in the search for a balance between tradition and progress, and between faith and reason, that Krausist ideals contributed to the cultural assimilation of industry.

By creating contrasts between past and present, the social foundries analyzed in this book offer valuable insights into the cultural repercussions of various phenomena related to industrial modernization, such as population displacement, environmental destruction, or labor exploitation. In this context, idealizations of the past embodied in representations of the rural space as idyllic and harmonic and of progress as chaotic and detrimental operate as a backdrop for the depiction of the moral and physical degradation of society. By describing the different actors and situations implied in this social transformation, political, journalistic, and literary texts managed to create detailed, though sometimes pessimistic, portraits of the industrial landscape. One can argue that the works studied here present industrial modernization as a violent process. This violence, manifested both physically (environmental disaster, social-class tensions) and symbolically (imposition of new forms of reasoning), had profound repercussions for how authors addressed social change—nostalgic feelings for the past, profound skepticism of the present, or pessimism about the future were thus common reactions that shaped their particular assessments of society.

In sum, the authors and ideas discussed in this book show how incompatible ideologies can produce a coherent narrative of national re-evaluation in light of scientific, technological, and industrial transformation. Multiple ways of understanding the world collided in the rise of industrial modernization—the emergence of the social foundry as an interpretative tool is just one of many attempts to make sense of the new cultural, social, and economic codifications of reality. While all the authors in this study were able to formulate or analyze social models using as a backdrop the conceptual and operational disposition of the foundry, none of them were actually involved in industrial enterprises. This paradox, however, highlights industry's capacity to produce its own scientific, anthropological, or sociological imagery, and to impose new cognitive frameworks. It also underscores authors' agency as intellectuals, namely as interpreters of reality and agents of social transformation. If the idea of nation that prevailed in nineteenth-century Spain was in part the result of political capitalization on different feelings toward tradition, it comes as no surprise that different, opposed reactions to industrial and technological change ended up radicalizing the tensions between progressive and conservative political ideologies. Tragically, this antagonism would expose its devastating potential in the multiple social confrontations that marked the twentieth century.

Acknowledgments

Completing a book is emotionally and mentally taxing. Academic writing is remarkably demanding, time consuming, and often unrewarding; much more so when it is attempted in a foreign language. When I look back at the time I have devoted to this endeavor, it seems impossible not to be indebted to a considerable amount of friends and colleagues that in one way or another contributed to its conclusion. Through the past years, and certainly amid the recent and daunting global challenges, however, it has been my family that has offered me the necessary stimulation to finalize this project. I am extremely grateful for the wholehearted motivation provided by my daughters, Violetta and Raphaella, and for the love, encouragement, and partnership of my wife, Allison, who has not only been my intellectual interlocutor but also a keen editor of my work in English.

Parts of this project were conceived, written, or reworked in New York, Madrid, Bogotá, Philadelphia, Collegeville, Mexico City, London, Lubbock, Dallas, Austin, and Madison. In all these places, I received motivation and help from too many people and institutions to properly name and thank. Special mention should be made, nonetheless, of my mother-in-law, Kim Heilbrun, for her wholehearted assistance in most of those cities over the years.

Embryonic versions of several sections in this book constituted my doctoral dissertation at Columbia University. Among the many colleagues and friends that contributed to bring that project to fruition, I would like to especially recognize my mentor, Wadda Rios-Font, whose timely advice and indefatigable support were pivotal to articulate some of the innovative concepts and original interpretations deployed in this study. I am also indebted to Carlos Alonso, Ronald Briggs, Eduardo Hernández Cano, Jo Labanyi, and Alberto Medina for questioning my analyses and highlighting the broader implications of my thesis.

I am also grateful for the generous support offered by Ursinus College. Work on this book was made possible thanks to a pre-tenure sabbatical and various summer research grants that allowed me to revisit the archives and focus on writing. I also want to thank my colleagues in the Modern Languages Department at Ursinus for their unconditional support and friendship. I am especially indebted to Teresa Ko for her mentorship and encouragement and for the insightful reading of several parts of this manuscript.

I also owe particular gratitude to the anonymous reviewers of the manuscript for their useful insights. Their recommendations were instrumental in strengthening the theoretical apparatus and overall structure of the book.

Finally, I would like to express my thanks to the editors of Campos Ibéricos at Bucknell University Press for their consideration of my project, and to Pamelia Dailey and Suzanne Guiod for their diligent guidance and assistance throughout the publishing process. I can hardly think about a better home for my book. I would be remiss if I did not also acknowledge here the professionalism and dedication of the editorial staff at Westchester Publishing Services, whose painstaking attention to detail was decisive in the flawless production of this volume.

Portions of chapter 3 have already appeared in print: the section on Zoel García de Galdeano draws from the article "The Science of Science Education: Zoel García de Galdeano, Mathematics Instruction, and the Progress of Spain," originally published in vol. 14, issue no. 1 (2017) of *Decimonónica, Revista de Producción Cultural Hispánica Decimonónica*; and the analysis of Laureano Calderón's ideas is based on the article "Laureano Calderón's Social Foundry: Industrial Imagination and Social Prescription in Restoration Spain," published in *Journal of Spanish Cultural Studies*, vol. 19, issue no. 1 (2019). I would like to thank *Decimonónica*'s managing editor and Taylor & Francis Group for their permission to reprint segments extracted from those pieces.

Notes

INTRODUCTION

1. See Walter Benjamin, "The Work of Art in the Age of Its Technological Reproducibility," in *The Work of Art in the Age of Its Technological Reproducibility, and Other Writings on Media*, ed. Brigid Doherty, Michael W. Jennings, and Thomas Y. Levin (Cambridge: Belknap Press of Harvard University Press, 2008), 22–23.

2. See, for instance, Stefan Pohl-Valero, "Termodinámica, pensamiento social y biopolítica en la España de la Restauración," *Universitas Humanística*, no. 69 (2010): 33–58; and Geraldine Lawless, *Modernity's Metonyms: Figuring Time in Nineteenth-Century Spanish Stories* (Lewisburg: Bucknell University Press, 2011). These studies approach nineteenth-century Spanish cultural production not as the result of the particular transformations of scientific and technological development but as symptomatic of a new way of thinking stimulated by such changes.

3. George Lakoff and Mark Johnson, *Metaphors We Live By* (Chicago: University of Chicago Press, 1980), 235.

4. In 2012, the *MIT Technology Review* and the Instituto Español de Comercio Exterior (Spanish Foreign Trade Agency) (ICEX) evaluated Spain's developments in terms of technology, dedicating articles to presenting the state of the country in different sectors of the economy. See "New Technologies from Spain," http://icex.technologyreview.com/ (accessed February 1, 2021).

5. Juan Carlos I de Borbón, "Palabras de Su Majestad el Rey en la inauguración de la nueva Línea de Alta Velocidad Madrid-Valencia," *Casa de S.M. el Rey* (Diciembre 18, 2010), https://www.casareal.es/EN/Actividades/Paginas/actividades_discursos_detalle.aspx?data=2482 (accessed February 1, 2021).

6. See "Líneas de Alta Velocidad" in the folder "Infraestructuras y Estaciones." Ministerio de Fomento de España, "Líneas de alta velocidad," http://www.adif.es/ (accessed February 1, 2021).

7. Although there is no consensus on the Restoration's chronology, many historians coincide in their assertion that the kind of political culture that prevailed from 1875 until the military coup of Miguel Primo de Rivera in 1923 was characteristic of the spirit of conciliation that helped overcome most of the ideological tensions during the second part

of the nineteenth century. See, for example, Gabriel Tortella, *El desarrollo de la España contemporánea: Historia económica de los siglos XIX y XX* (Madrid: Alianza, 1998); Juan Pablo Fusi, *España: La evolución de la identidad nacional* (Madrid: Temas de hoy, 2000); and José Álvarez Junco, *Mater dolorosa: La idea de España en el siglo XIX* (Madrid: Taurus, 2003).

8. Juan Carlos I de Borbón, "Palabras de Su Majestad el Rey."

9. Ramón Torres Muñoz de Luna, "Las dos Isabeles: Composición poética para solemnizar la inauguración del ferro-carril de León a Valencia," in Biblioteca Nacional, Madrid, MSS/12944/103 (1863).

10. See, in particular, Fusi, *España*, 155–162; and Álvarez Junco, *Mater dolorosa*, 129–134.

11. See Álvarez Junco, *Mater dolorosa*, 194. Tracing the influential role that the Catholic clergy played in the context of liberal reform, Gregorio Alonso has also shown the relationship that existed between politics and religion in the context of modernization. See Gregorio Alonso, "How to Be Religious under Liberalism," in *Spain in the Nineteenth Century: New Essays on Experiences of Culture and Society*, ed. Andrew Ginger and Geraldine Lawless (Manchester: Manchester University Press, 2018), 89–108.

12. Fusi, *España*, 184.

13. Francisco del Río Joan, "La electricidad en España al advenimiento de Don Alfonso XIII," in *La ciencia y la industria eléctrica en España al subir al trono S.M. el Rey Don Alfonso XIII* (Madrid: La Energía Eléctrica, 1902), 26.

14. For detailed accounts on regenerationism, see Joseph Harrison, "The Regenerationist Movement in Spain after the Disaster of 1898," *European History Quarterly* 9, no. 1 (1979): 1–27; and Joseph Harrison and Alan Hoyle, eds., *Spain's 1898 Crisis: Regenerationism, Modernism, Post-Colonialism* (Manchester: Manchester University Press, 2000), 1–51.

15. Transformations prompted by industrialization during the nineteenth century shaped social thinking and the way social disciplines approached politics, economy, and culture. The consolidation of science and technology as essential tools for the advancement of society also problematized the way in which transitions between artisanal and mass production, or between notions of *production for use* and *production for profit* were understood. Different thinkers contributed to the evaluation of these changes with conceptual, ideological, and philosophical apparatuses that offered insight into progress's new social configurations and the tensions generated at the encounter of modernity and traditional values, religious dogmas, and nationalistic feelings. Auguste Comte or Karl Marx would set the basis in that regard for what later Émile Durkheim, Max Weber, or Georg Simmel would champion as sociology of capitalism. See Lewis A. Coser, introduction to *The Division of Labor in Society*, by Emile Durkheim, trans. W. D. Halls (New York: Free Press, 1984), xii–xiii. Introduced, translated, or promulgated by Spanish authors after their readings of Friedrich Nietzsche, Arthur Schopenhauer, or Allan Kardec, philosophical currents such as voluntarism, iconoclasm, and spiritualism also concurred at the turn of the twentieth century in Spain. See Michael Candelaria, *The Revolt of Unreason: Miguel de Unamuno and Antonio Caso on the Crisis of Modernity* (Amsterdam: Rodopi, 2012), 3.

16. See Candelaria, *The Revolt of Unreason*, 14.

17. See Ian Hacking, *The Taming of Chance* (Cambridge: Cambridge University Press, 1990), 5, 39; and Richard Olson, *Science and Scientism in 19th-Century Europe* (Urbana: University of Illinois Press, 2008), 2, 21, 87, 250ff.

18. Agustí Nieto-Galan, "A Republican Natural History in Spain around 1900: Odón de Buen (1863–1945) and His Audiences," *Historical Studies in the Natural Sciences* 42, no. 3 (2012): 161–164.

19. Juan López-Morillas, *El krausismo español: Perfil de una aventura intellectual* (México: Fondo de Cultura Económica, 1980), 13.

20. See López-Morillas, *El krausismo*, 9.

21. Michael Aronna, *"Pueblos enfermos": The Discourse of Illness in the Turn-of-the-Century Spanish and Latin American Essay* (Chapel Hill: University of North Carolina Press, 1999), 14.

22. Julián Sanz del Río, "Cartas inéditas," in *Revista Europea* 1, no. 3 (1874): 69, quoted in *El krausismo español*, 95.

23. Stefan Pohl-Valero, "The 'Circulation' of Energy: Thermodynamics, National Culture, and Social Progress in Spain, 1868–1890," in *Popularizing Science and Technology in the European Periphery, 1800–2000*, ed. Faidra Papanelopoulou, Agustí Nieto-Galán, and Enrique Perdiguero (Burlington: Ashgate, 2009), 130.

24. As Gonzalo Capellán de Miguel explains, Orovio's reform not only pushed the consolidation of the Krausists' educational project through the creation of the Institución Libre de Enseñanza but also prepared the ground for a new pedagogy oriented toward the strengthening of elementary and secondary schooling. One of the most active members in the development of methodologies and contents to fulfill this mission was Giner de los Ríos's protégé, Manuel Bartolomé Cossío (1875–1935). Through his labor in the Museo Pedagógico Nacional, Cossío contributed to the creation of the Colonias Escolares de Vacaciones (kids' summer camps) in 1887, spaces dedicated to the exploration of nature and open to children of all social backgrounds. See Gonzalo Capellán de Miguel, *De Orovio a Cossío: Vieja y nueva educación, La Rioja (1833–1933)* (Logroño: Instituto de Estudios Riojanos, 1999), 72ff.

25. See Sandie Eleanor Holguin, *Creating Spaniards: Culture and National Identity in Republican Spain* (Madison: University of Wisconsin Press, 2002), 144.

26. See Roberta Johnson, *Crossfire: Philosophy and the Novel in Spain, 1900–1934* (Lexington: University Press of Kentucky, 1993), 16–17.

27. Antonio Cánovas del Castillo, *Discurso pronunciado el día 6 de noviembre de 1882 en el Ateneo Científico y Literario de Madrid con motivo de la apertura de sus cátedras* (Madrid: Imprenta Central a cargo de Víctor Saiz, 1882), 66 (my emphasis).

28. See Nicole Boivin, *Material Cultures, Material Minds: The Impact of Things on Human Thought, Society, and Evolution* (Cambridge: Cambridge University Press, 2008); and Ismael Sarmiento Ramírez, "Cultura y cultura material: Aproximaciones a los conceptos e inventario epistemológico," *Anales del Museo de América*, no. 15 (2007): 217–236.

29. See Ian Woodward, *Understanding Material Culture* (London: Sage, 2007).

30. See Jordi Maluquer de Motes, "Factores y condicionamientos del proceso de industrialización en el siglo XIX: El caso español," in *La industrialización del norte de España*, ed. Jordi Maluquer de Motes, Emiliano Fernández de Pinedo, and José Luis Hernández Marco (Barcelona: Crítica, 1988), 13–36.

31. See Antonio Gómez Mendoza, *Ferrocarril, industria y mercado en la modernización de España* (Madrid: Espasa Calpe, 1989); and Tortella, *El desarrollo de la España contemporánea*.

32. See Stephanie Anne Sieburth, *Inventing High and Low: Literature, Mass Culture, and Uneven Modernity in Spain* (Durham: Duke University Press, 1994); Jo Labanyi, *Gender and Modernization in the Spanish Realist Novel* (Oxford: Oxford University Press, 2000); Francisco Caudet, *El parto de la modernidad: La novela española de los siglos XIX y XX* (Madrid: De la Torre, 2002); and Jesús Cruz, *The Rise of Middle-Class Culture in Nineteenth-Century Spain* (Baton Rouge: Louisiana State University Press, 2011).

33. See, among others, Lily Litvak, *Transformación industrial y literatura en España (1895–1905)* (Madrid: Taurus, 1980); Juan Carlos Ponce, *Literatura y ferrocarril en España: Aspectos socioliterarios del ferrocarril en España* (Madrid: Fundación de los Ferrocarriles Españoles, 1996); Juan Cano Ballesta, *Literatura y tecnología: Las letras españolas ante la Revolución Industrial, 1890–1940* (Valencia: Pre-Textos, 1999); José A. López Cerezo and José M. Sánchez Ron, *Ciencia, tecnología, sociedad y cultura en el cambio de siglo* (Madrid: Biblioteca Nueva, 2001); Dale J. Pratt, *Signs of Science: Literature, Science, and Spanish Modernity since 1868* (West Lafayette: Purdue University Press, 2001); Benigno Delmiro Coto, *Literatura y minas en la España de los siglos XIX y XX* (Gijón: Trea, 2003); Santos Casado de Otaola, *Naturaleza patria: Ciencia y sentimiento de la naturaleza en la España del regeneracionismo* (Madrid: Marcial Pons, 2010); and Ryan A. Davis and Alicia Cerezo Paredes, eds., *Modernity and Epistemology in Nineteenth-Century Spain: Fringe Discourses* (Lanham: Lexington Books, 2017).

CHAPTER 1 — THE SOCIAL FOUNDRY

1. Antonio Lafuente and Tiago Saraiva suggest that the transformation of Madrid, from being the seat of the Royal Court to becoming the capital of a modern state, is directly related to the processes of industrial development. For the authors, Madrid turned into a laboratory wherein the effects of material modernization on society acted as the ideological reference point of the city's urban projection: "[Urban planning] was a public debate, affecting all aspects of city life, from drains to transport, and from public and private hygiene to the layout and naming of streets." See Antonio Lafuente and Tiago Saraiva, "The Urban Scale of Science and the Enlargement of Madrid (1851–1936)," *Social Studies of Science* 34, no. 4 (2004): 532.

2. Ángel Fernández de los Ríos, *Guía de Madrid, manual del madrileño y del forastero* (Madrid: Oficinas de La Ilustración Española y Americana, 1876), 679.

3. Bruce Clarke, *Energy Forms: Allegory and Science in the Era of Classical Thermodynamics* (Ann Arbor: University of Michigan Press, 2001), 5.

4. See Laura Otis, *Networking: Communicating with Bodies and Machines in the Nineteenth Century* (Ann Arbor: University of Michigan Press, 2001); and "The Metaphoric Circuit: Organic and Technological Communication in the Nineteenth Century," *Journal of the History of Ideas* 63, no. 1 (2002): 105–128; Philip Mirowski, *More Heat Than Light: Economics as Social Physics, Physics as Nature's Economics* (Cambridge: Cambridge University Press, 1989); Nicholas Daly, *Literature, Technology, and Modernity, 1860–2000* (Cambridge: Cambridge University Press, 2004); Ted Underwood, *The Work of the Sun: Literature, Science, and Political Economy, 1760–1860* (New York: Palgrave Macmillan, 2005); Barri Gold, *ThermoPoetics: Energy in Victorian Literature and Science* (Cambridge: MIT, 2010); and Tamara Ketabgian, *The Lives of Machines: The Industrial Imaginary in Victorian Literature and Culture* (Ann Arbor: University of Michigan Press, 2011).

5. See José Manuel Sánchez Ron, *Cincel, martillo y piedra: Historia de la ciencia en España (siglos XIX y XX)* (Madrid: Taurus, 2000); and Stefan Pohl-Valero, "The

'Circulation' of Energy: Thermodynamics, National Culture, and Social Progress in Spain, 1868–1890," in *Popularizing Science and Technology in the European Periphery, 1800–2000*, ed. Faidra Papanelopoulou, Agustí Nieto-Galán, and Enrique Perdiguero (Burlington: Ashgate, 2009), 115–134; "La comunicación de la termodinámica. Física, cultura y poder en la España de la segunda mitad del siglo XIX," *Memoria y Sociedad* 13, no. 27 (2009): 121–141; "Termodinámica, pensamiento social y biopolítica en la España de la Restauración," *Universitas Humanística*, no. 69 (2010): 33–58; and *Energía y cultura: Historia de la termodinámica en la España de la segunda mitad del siglo XIX* (Bogotá, D.C.: Pontificia Universidad Javeriana, 2011).

6. See Pohl-Valero, "La comunicación de la termodinámica," 134.

7. For a detailed analysis of the political and social uses of Joule's conceptualization of heat and the laws of thermodynamics in Restoration Spain, see Pohl-Valero, "The 'Circulation' of Energy," "La comunicación de la termodinámica," "Termodinámica, pensamiento social y biopolítica," and *Energía y cultura*. For approaches to the emergence of Darwinism and its dialogue with peninsular cultural production, see, for example, Thomas F. Glick, *The Comparative Reception of Darwinism* (Chicago: University of Chicago Press, 1988); Dale J. Pratt, *Signs of Science: Literature, Science, and Spanish Modernity since 1868* (West Lafayette: Purdue University Press, 2001); and "'Fundamental Truths' to 'Playthings of Science': Science, Technology, and Modernity in Pardo Bazán," in *Approaches to Teaching the Writings of Emilia Pardo Bazán*, ed. Margot Versteeg and Susan Walter (New York: Modern Language Association of America, 2017), 71–77; and Travis Landry, *Subversive Seduction: Darwin, Sexual Selection, and the Spanish Novel* (Seattle: University of Washington Press, 2012).

8. The philosophical foundations of this dialogue were initially postulated in the pioneering work of Michel Serres. From his perspective, the encounter of the epistemological structures of science with the rhetoric of social discourse elicited multiple semiotic transformations that reconfigured the whole apparatus of representation. See Michel Serres, *Hermes—Literature, Science, Philosophy*, ed. David F. Bell and Josué V. Harari (Baltimore: Johns Hopkins University Press, 1982), 44–45.

9. Greg Myers, "Nineteenth-Century Popularizations of Thermodynamics and the Rhetoric of Social Prophecy," *Victorian Studies* 29, no. 1 (1985): 49.

10. George Lakoff and Mark Johnson, *Metaphors We Live By* (Chicago: University of Chicago Press, 1980), 157.

11. Christopher Y. Tilley, *Metaphor and Material Culture* (Oxford: Blackwell, 2000), 4.

12. Tilley, *Metaphor*, 8.

13. Otis, "The Metaphoric Circuit," 127–128.

14. Lakoff and Johnson offer a detailed description and analysis of this process of cataloging. See *Metaphors*, 25–32.

15. Clarke, *Energy Forms*, 62.

16. Anson Rabinbach, *The Human Motor: Energy, Fatigue, and the Origins of Modernity* (New York: BasicBooks, 1990), 14.

17. Rom Harré, "Material Objects in Social Worlds," *Theory, Culture & Society* 19, no. 5–6 (2002): 32.

18. Clarke, *Energy Forms*, 20 (emphasis in original).

19. Rabinbach, *The Human Motor*, 3.

20. Gilles Deleuze, *Difference and Repetition*, trans. Paul Patton (New York: Columbia University Press, 1994), 1.

21. Deleuze, *Difference and Repetition*, 28.

22. See Arjun Appadurai, *The Social Life of Things: Commodities in Cultural Perspective* (Cambridge: Cambridge University Press, 2001).

23. Ernest Gellner points precisely to the social and political reorganization that the consolidation of industrial economies produced in Europe. He proposes that the convergence of culturally diverse groups in urban and industrial centers demanded the consolidation of collective identifiers to facilitate mass production. The idea of a group of people working together toward a common goal was then the germ of the modern nation. In that process, science and technology legitimized the social and political mechanisms put into effect to disseminate particular visions of the past, present, and future of the country. See Ernest Gellner, *Nations and Nationalism* (Ithaca: Cornell University Press, 1983), 19–37.

24. Bourdieu's cultural field is a symbolic space wherein the production of meaning is regulated by the specific disposition of power relations and ideological constructs: "The space of literary or artistic position-takings, i.e. the structure set of the manifestations of the social agents involved in the field—literary or artistic works, of course, but also *political acts of pronouncement, manifestos or polemics, etc.*—is inseparable from the space of literary or artistic positions defined by possession of a determinate quantity of specific capital (recognition) and, at the same time, by occupation of a determinate position in the structure of the distribution of this specific capital. The literary or artistic field is a field of forces, but it is also a field of struggles tending to transform or conserve this field of forces." See Pierre Bourdieu, *The Field of Cultural Production: Essays on Art and Literature*, ed. Randal Johnson (Cambridge: Polity, 1999), 30 (my emphasis).

25. Harré, "Material Objects," 23.

26. Michel Foucault, *The Archaeology of Knowledge; The Discourse on Language*, trans. Rupert Swyer (New York: Pantheon, 1972), 42.

27. To refer to this range of relations, Foucault uses the term episteme (*épistémè*): "By episteme, we mean, in fact, the total set of relations that unite, at a given period, the discursive practices that give rise to epistemological figures, sciences, and possibly formalized systems; . . . The episteme is not a form of knowledge (*connaissance*) or type of rationality which, crossing the boundaries of the most varied sciences, manifests the sovereign unity of a subject, a spirit, or a period; it is the group of relations that can be discovered, for a given period, between the sciences when one analyses them at the level of discursive regularities." See Foucault, *Archaeology*, 211.

28. Foucault, *Archaeology*, 122.

29. Foucault, *Archaeology*, 139.

30. Foucault criticizes approaches to cultural production wherein discourse is just seen as an allegorical representation of social, economic, and political realities: "It is in order to be sure that this occurrence [the text's] is not linked with synthesizing operations of a purely psychological kind (the intention of the author, the form of his mind, the rigour of his thought, the themes that obsess him, the project that traverses his existence and gives it meaning), and to be able to grasp other forms of regularity, other types of relations. Relations between statements (even if the author is unaware of them; even if the statements do not have the same author; even if the authors were unaware of each other's existence); relations between groups of statements thus established (even if these groups do not concern the same, or even adjacent, fields; even if they do not possess the same formal level; even if they are not the locus of assignable exchanges); relations

between statements and groups of statements and events of a quite different kind (technical, economic, social, political)." See Foucault, *Archaeology*, 28–29.

31. Michel Serres, *Hermes IV: La Distribution* (Paris: Minuit, 1977), 42–43.

32. See Georg Simmel, *Simmel on Culture: Selected Writings*, ed. David Frisby and Mike Featherstone (London: Sage, 2000), 118.

33. An example of this is Frederick Winslow Taylor's method for classifying movements and measuring time in assembly lines. Between 1890 and 1893, Taylor was the manager of a paper mill, where he developed a tool to calculate the time spent in each process in the factory; he was thus able to estimate the efficiency of production and the mechanisms to improve it. The method Taylor implemented was very influential in the organization of American industrial spaces at the beginning of the twentieth century and remains part of many current manufacturing settings around the world.

34. Aram A. Yengoyan, Rev. of *Simmel on Culture Selected Writings*, ed. David Frisby and Mike Featherstone, *Comparative Studies in Society and History* 44, no. 3 (2002): 621.

35. Simmel, *Simmel on Culture*, 174–186.

36. Nicole Boivin, *Material Cultures, Material Minds: The Impact of Things on Human Thought, Society, and Evolution* (Cambridge: Cambridge University Press), 8–9.

37. Michel de Certeau, *The Practice of Everyday Life*, trans. Steven Rendall (Berkeley: University of California Press, 2002), 113.

38. In the context of the urban expansion that followed industrialization, the question of social control gained relevance. Among the systems developed to discipline the growing prison population in Europe, Jeremy Bentham's became widely used. The panopticon, as Bentham denominated this system, was designed to allow a centralized and constant supervision of prisoners by the strategic distribution of cells and the precise location of surveillance posts. This conceptual model re-emerges throughout modern industrial contexts to generate a self-regulated society wherein people have internalized surveillance and function accordingly. See Michel Foucault, *Discipline and Punish: The Birth of the Prison*, trans. Alan Sheridan (New York: Vintage Books, 2012), 224.

39. Matthew Beaumont and Michael J. Freeman, *The Railway and Modernity: Time, Space, and the Machine Ensemble* (Oxford: Peter Lang, 2007), 47.

40. Pedro Antonio de Alarcón, *Viajes por España* (Madrid: Imprenta de A. Pérez Durrul, 1883), 286 (unfinished in the original).

41. Alarcón, *Viajes por España*, 289.

42. Alarcón, *Viajes por España*, 286.

43. Benito Pérez Galdós, "La novela en el tranvía," in *13 cuentos*, ed. Esteban Gutiérrez Díaz-Bernardo (Madrid: Edaf, 2001), 176.

44. Stephen Kern, *The Culture of Time and Space, 1880–1918* (Cambridge: Harvard University Press, 2003), 21–22.

45. The musical scene in fin-de-siglo Spain was transformed after industrialization. This period coincided with the emergence of crucial figures such as composers Isaac Albéniz (1860–1909), Enrique Granados (1867–1916), and Manuel de Falla (1876–1946). In these musicians' melodic style, it is possible to perceive a particular understanding of the processes of massification or the acceleration of society. In general terms, their work represented the peak of a renaissance that can be associated ideologically with the Generación del 98, and stylistically with impressionism, as it was directly influenced by French and Russian musicians experimenting with notions of time and space through linearity, tempo, or tonal aspects. See José Gregorio Cayuela Fernández, *Un siglo de*

España: Centenario 1898–1998 (Ciudad Real: Universidad de Castilla La Mancha, 1998). Similarly, the invention of the cinematograph plays with visual perception by dynamically projecting fixed images. The railroad was perceived to be a source of inspiration for the first movies, and both in France and Spain the first film experiments included images of trains. Such is the case of *L'arrivée d'un train en gare de La Ciotat* (1895) by Auguste and Louis Lumière, and *Choque de trenes* (1902) by Segundo de Chomón.

46. See Stephanie Anne Sieburth, *Inventing High and Low: Literature, Mass Culture, and Uneven Modernity in Spain* (Durham: Duke University Press, 1994).

47. Advances in the study of electricity and magnetism reached a critical point with the discovery of magnetic induction by Michael Faraday in 1831. The concept of field derived from these scientific advances changed the way in which the relation between objects and their surroundings was understood, facilitating developments such as the theorization of relativity in the early twentieth century. See N. Katherine Hayles, *Chaos Bound: Orderly Disorder in Contemporary Literature and Science* (Ithaca: Cornell University Press, 1990), xi–xii.

48. Thomas Newcomen's invention of the first steam engine in 1710 was not only a mechanical achievement, it also entailed important scientific advances that included, among others, Robert Boyle's understanding of the connection between heat, pressure, and volume, and Nicolas Carnot's theorization of heat dissipation. While Boyle's work opened the possibility of creating devices that manipulated volume, temperature, and pressure, Carnot's postulations explained the circulation of energy within a steam engine—what later came to be known as Carnot cycle. Setting the principles of equilibrium and dissipation of heat, these studies were also essential for the development of the laws of thermodynamics. See Gold, *ThermoPoetics*, 58–59.

49. Pohl-Valero, "La comunicación de la termodinámica," 126.

50. Jo Labanyi, *Gender and Modernization in the Spanish Realist Novel* (Oxford: Oxford University Press, 2000), 63.

51. Pohl-Valero, "Termodinámica, pensamiento social y biopolítica," 42.

52. Rabinbach provides an enlightened description of how the worker–machine association was conceived under these premises. See *The Human Motor*, 55ff. In this regard, Pohl-Valero also offers a complete historical analysis to understand the adoption of similar ideas in the Spanish case. See "Termodinámica, pensamiento social y biopolítica," 44.

53. Ludwig Büchner, *Fuerza y materia: Estudios populares de historia y filosofía naturales*, trans. A. Avilés (Madrid: Librería de Fernando Fe, 1878), 65 (my emphasis).

54. Pohl-Valero, "La comunicación de la termodinámica," 126, 133.

55. Enrique Serrano Fatigati, "La evolución en la naturaleza," *Revista de la Universidad de Madrid*, no. 3 (1874): 511.

56. Serrano Fatigati, "La evolución," 292–296.

57. See José Rodríguez Mourelo, "Concepto de la energía. Curso de ciencias naturales: Tercera conferencia, 12 de enero de 1882," *Revista Contemporánea* 8, no. 40 (1882): 275–311.

58. Rodríguez Mourelo, "Concepto de la energía," 290.

59. José Echegaray, "Aplicación de las fuerzas naturales á la industria y al comercio," *Conferencias del curso de 1879 á 1880. Círculo de la Unión Mercantil* (Madrid: Imprenta y estereotipia de El Liberal, 1881), 106.

60. Echegaray, "Aplicación de las fuerzas," 107 (emphasis in original).

61. It must be noted that industrialization in Spain did not consolidate in a uniform way across the country. Some regions experienced a boom in their economies due to a large extent to foreign investment. This injection of capital materialized in the development of infrastructure (railroads, factories, industrial mining sites, and so forth) and the displacement of populations to supply the demand for workers. Areas where heavy industry developed also saw the deterioration in the living conditions of both locals and newcomers as social and labor tensions worsened. In many cases, these frictions evolved into forms of violence with nationalistic undertones. See Juan Pablo Fusi, *Política obrera en el país vasco: 1880–1923* (Madrid: Turner, 1975); Jordi Maluquer de Motes, "Factores y condicionamientos del proceso de industrialización en el siglo XIX: El caso español," in *La industrialización del norte de España*, ed. Jordi Maluquer de Motes, Emiliano Fernández de Pinedo, and José Luis Hernández Marco (Barcelona: Crítica, 1988), 13–36; Sima Lieberman, *The Contemporary Spanish Economy: A Historical Perspective* (London: Routledge, 2006), 120–199; Miguel Á. Pérez de Perceval Verde and Alejandro Sánchez Rodríguez, "El capital extranjero en el desarrollo de la minería española del siglo XIX," in *La inversión extranjera en la minería española*, ed. Albert Broder et al. (Madrid: Instituto Geológico y Minero de España, 2014), 7–69; and Leandro Prados de la Escosura, "Foreign Capital in 19th Century Spain's Investment Boom," *European Review of Economic History* 24, no. 2 (2020): 314–331.

62. After Bessemer's invention, in 1857 Sir Carl Wilhelm Siemens developed a system that regenerated heat and saved important amounts of combustible material during the smelting process. Using those improvements as the basis of his work, in 1865 Pierre-Émile Martin further improved the efficiency of the system by using an open core in which the metal oxidation contributed to the total heat of the process. This process became known as the Siemens-Martin conversion, and by the end of the century had replaced most of the Bessemer furnaces. For detailed descriptions of the importance of these technological advancements, see Eric J. Hobsbawm, *Industry and Empire, from 1750 to the Present Day* (New York: The New Press, 1999), 159ff.

63. The Constitution of 1812 has been recognized as one of the most advanced documents of its kind. This was somewhat contradictory, as José Álvarez Junco underscores, given that not many years before, Spain was associated with everything the Constitution wanted to abolish: absolutism, intolerance, and dogmatism. The document promulgated in Cádiz considered the abolition of the feudal system, the guarantee of individual freedoms, the recognition of personal and judicial rights, the eradication of torture, the establishment of statutes to confiscate uncultivated Church property, the declaration of national sovereignty, and the division of the state into three branches: legislative, executive, and judicial. See Jose Álvarez Junco, *Mater dolorosa: La idea de España en el siglo XIX* (Madrid: Taurus, 2003), 117.

64. See Sánchez Ron, *Cincel, martillo y piedra*, 42.

65. See, for example, Álvarez Junco, *Mater dolorosa*, 361–362.

66. See Gonzalo Capellán de Miguel, *De Orovio a Cossío: Vieja y nueva educación, La Rioja (1833–1933)* (Logroño: Instituto de Estudios Riojanos, 1999), 36–40.

67. Jonathan Smith, *Fact and Feeling: Baconian Science and the Nineteenth-Century Literary Imagination* (Madison: University of Wisconsin Press, 1994), 10.

68. Scientific backwardness in fin-de-siglo Spain was commonly attributed to three factors: religious devotion, deficiencies in higher education, and lack of industrial initiatives. In contrast to countries like England or France, original scientific developments

were scarce on the peninsula. Referring to disciplines like physics or chemistry, Sánchez Ron adds: "Aun en el caso de que en España se hubiesen superado con creces los vicios de la educación científica que se encuentran en otras naciones que contribuyeron de manera apreciable a la ciencia del siglo XIX, la diferencia entre los logros de las ciencias físico-químicas en España y en otras naciones es, comparativamente, mucho mayor que la correspondiente diferencia entre los 'vicios' de los respectivos sistemas educativos" (Even in the case that Spain had been able to dramatically correct its educational problems and approach the level of other countries who actively contributed to the nineteenth-century scientific progress, the difference in the development of the physical and chemical sciences between Spain and other nations was, comparatively, far bigger than the one between the "vices" of their respective systems). See José Manuel Sánchez Ron, "Las ciencias físico-matemáticas en la España del siglo XIX," *Ayer*, no. 7 (1992): 53. This did not imply, however, the complete absence of original scientific and technological work, neither at the turn of the twentieth century nor earlier. Although official narratives often echoed other countries' or conservative sectors' political and economic interests in perpetuating the Black Legend of Spanish backwardness, there were prominent scientists and innovators in Spain (some of them studied in this book) who actively contributed to fields such as chemistry, physics, mathematics, urban engineering, and military technology. They indeed belong to a long tradition of scientific and technological development that can be traced back to the sixteenth century, and that has been ignored or dismissed historically for political reasons. See Jorge Cañizares-Esguerra, "Iberian Science in the Renaissance: Ignored How Much Longer?" *Perspectives on Science* 12, no. 1 (2004): 86–124; and *Nature, Empire, and Nation: Explorations of the History of Science in the Iberian World* (Palo Alto: Stanford University Press, 2006), 45, 96ff.

69. See, for example, J. D. Bernal, *Science and Industry in the Nineteenth Century* (London: Routledge, 2011), 142–144.

70. Classic studies on the problematic industrial development in Spain account for the numerous obstacles that the country experienced in its transition from an agrarian economy to a system based on the large-scale textile, mining, and steel production. See, for example, Jordi Nadal, *El fracaso de la revolución industrial en España, 1814–1913* (Barcelona: Ariel, 1977).

71. José Echegaray, "Discurso leído por José Echegaray ante la Real Academia de Ciencias Exactas, Físicas y Naturales en su recepción pública el 11 de marzo de 1866." *Revista de Obras Públicas*, no. 14 (1866): 93.

72. Echegaray, "Discurso Real Academia de Ciencias," 100.

73. Echegaray, "Discurso Real Academia de Ciencias," 101.

74. As Sánchez Ron explains, the polemic sparked by José Echegaray's speech would have further repercussions on legislative debates proposing the creation of institutions dedicated to support the development of science in the country. The polemic reached its peak after historian Gaspar Núñez de Arce delivered his speech "Causa de la precipitada decadencia y total ruina de la literatura nacional bajo los últimos reinados de la casa de Austria" (Causes for the precipitous decadence and total ruin of national literature under the late reigns of the House of Habsburg) during his induction to the Real Academia Española (Royal Spanish Academy) in 1874. Manuel de la Revilla's review of this speech would generate the response of conservative thinkers such as Marcelino Menéndez Pelayo, who did not agree with the idea of religious tradition being an obstacle for scientific advancement. Menéndez Pelayo contested Núñez de Arce's and Revilla's points

in "Mr. Masson redivivo" (Mr. Masson revived), an article published in the 1876 June issue of the journal *Revista Europea*; Menéndez Pelayo later expanded the article's main ideas in *Polémicas, indicaciones y proyectos sobre la ciencia española* (Debates, evidence, and projections on Spanish science) (1876) and *La ciencia española* (Spanish science) (1887–80). See Sánchez Ron, *Cincel, martillo y piedra*, 130–133. But the origin of the debate harks back to the late eighteenth century. In fact, the reference to Masson de Morvilliers pins the dispute's roots down to 1783, when, in the entry about Spain written for Diderot's *Encyclopédie Méthodique* (Methodical encyclopedia), the French thinker questioned the extent to which the country had actually contributed to European modernity. On the peninsula, reactions to this article exposed the existence of two different approaches to political and economic matters, either supporting or dismissing the need for institutional transformation. See Ernesto García Camarero, *La polémica de la ciencia española* (Madrid: Alianza, 1970); Ernesto García Camarero and Enrique García Camarero, *La ciencia española entre la polémica y el exilio: Dentro de la evolución de la ciencia europea* (Madrid: Megustaescribir, 2016), 59–68; and Cañizares-Esguerra, *Nature, Empire, and Nation*, 106–110.

75. In the Restoration, *integrismo* emerged as a form of antiliberalism that ecclesiastical authorities and conservative sectors of society promoted to counteract many of the anticlerical measures put in place during the revolutionary period (1868–1874). In 1875, Antonio Cánovas del Castillo, serving as president of the Consejo de Ministros (Cabinet of Ministers), had no choice but to establish alliances with these traditionalist political fronts in order to sustain the moderate system that was the basis of his political project. The appointment of Orovio as secretary of education was part of this conciliatory strategy. See Gonzalo Capellán de Miguel, "Política educativa bajo los gobiernos de Cánovas y Sagasta: Propuestas para una interpretación." *Berceo. Revista Riojana de Ciencias Sociales y Humanidades*, no. 139 (2000): 126–127.

76. Foreign investment was determining in the Spanish economic boom of the 1860s. This was in part the result of a flexible legislation and the tacit approval of the country's financial elite. See Prados de la Escosura, "Foreign Capital," 314; and Sima Lieberman, *The Contemporary Spanish Economy*, 120–121. This external interference, as I show in the following chapters, would provoke mixed reactions among thinkers and writers, who consistently criticize or lament the expropriation of natural resources and the exploitative conditions imposed on local workers.

77. M. K. Flynn, *Ideology, Mobilization, and the Nation: The Rise of Irish, Basque, and Carlist Nationalist Movements in the Nineteenth and Early Twentieth Centuries* (New York: Macmillan, 2000), 121.

78. It is worth noting that this dichotomic division stems from a generalization. In most cases there was not a clearly defined ideological separation among thinkers—many conservatives embraced the changes and challenges of modernization while many liberals showed strong attachments to the past and tradition. See Leoncio López Ocón, *Breve historia de la ciencia Española* (Madrid: Alianza, 2003), 11.

79. See Juan Pablo Fusi, "El legado del 98," in *Imágenes y ensayos del 98*, ed. Raymond Carr et al. (Valencia: Fundación Cañada Blanch, 1998), 291.

CHAPTER 2 — ECONOMY AND OTHER MATTERS OF STATE

1. Spain's geological map was an ongoing state project in the second half of the century convened to produce a detailed study of the entire peninsular territory. From the creation

in 1849 of the Instituto Geológico y Minero de España (Spain's Geological and Mining Institute), the project was assigned to the most highly recognized scientist of the period. In the beginning, the map was under the supervision of the Comisión para la Carta Geológica de Madrid y General del Reino (Commission for the Geological Map of Madrid and the General Kingdom), a state agency created by Juan Bravo Murillo to oversee research projects related to national geography. Around 1873, however, the subgroup working on the geological aspect of the commission had developed its own administrative organization, a unity that later came to be known as the Comisión del Mapa Geológico de España. The publication of the results of this research project in the *Boletín de la Comisión del Mapa Geológico de España* (Bulletin of the Commission of the Geological Map of Spain) appeared for the first time on June 30, 1873, and its release coincided with similar publications of geological societies in London and Paris. Scientists such as Lucas Mallada or Manuel Fernández de Castro were among the directors of the Commission during this late period. See José Manuel Sánchez Ron, *Cincel, martillo y piedra: Historia de la ciencia en España (siglos XIX y XX)* (Madrid: Taurus, 2000), 54–56; and Isabel Rábano, *Los cimientos de la Geología: la Comisión del Mapa Geológico de España (1849–1910)* (Madrid: Instituto Geológico y Minero de España, 2015).

2. Bruce Clarke, *Energy Forms: Allegory and Science in the Era of Classical Thermodynamics* (University of Michigan Press, 2001), 44.

3. José Echegaray, "Discurso leído por José Echegaray ante la Real Academia de Ciencias Exactas, Físicas y Naturales en su recepción pública el 11 de marzo de 1866," *Revista de Obras Públicas*, no. 14 (1866): 102.

4. In mathematics, Echegaray's studies deal with calculus and geometry, and some of his lectures were compiled as textbooks for the study of those areas in secondary- and college-level courses, as in the cases of *Cálculo de variaciones* (Multivariable calculus) (1858), *Problemas de geometría plana* (Plane geometry problems) (1865), *Problemas de geometría analítica* (Analytic geometry problems) (1865), and *Introducción a la geometría superior* (Introduction to advanced geometry) (1867). In physics, he published, among others, *Teorías modernas de la física* (Modern theories of physics) (1867), and *Tratado elemental de termodinámica* (Basic thermodynamics) (1868).

5. Among the more than 60 works published in this period, it is important to highlight pieces like *La esposa del vengador* (The bride of the avenger) (1874), *En el puño de la espada* (At the sword's point) (1875), *Locura o santidad* (Madman or saint?) (1876), *El gran Galeoto* (The great go-between) (1881), *Mariana* (1892), *Mancha que limpia* (The cleansing blotch) (1895), and *El loco Dios* (*The Madman Divine*) (1900). In many of these works, Echegaray attempted a complete formal and thematic renovation of drama to make it as popular as it was in the seventeenth century. This enormous production was possible in part due to Echegaray's ideological incompatibilities with the Restoration's agenda, which pushed him out of politics. He had modest roles, nonetheless, in the Government of Práxedes Sagasta, the constitution of the Institución Libre de Enseñanza in 1876, and the creation of the Republican Progressive Party in 1880. See Javier Fornieles Alcaraz, *Trayectoria de un intelectual de la Restauración: José Echegaray* (Almería: Cajalmería, 1989); and David Thatcher Gies, *The Theatre in Nineteenth-Century Spain* (Cambridge: Cambridge University Press, 1994), 295.

6. See, for example, Enrique and Natalia Pérez-Galdós, "La dualidad Echegaray," *Ingeniería y Territorio*, no. 79 (2007): 4–11.

7. Echegaray, "Discurso Real Academia de Ciencias," 98.

8. Despite its liberal tone and scientific orientation, Echegaray's literary work has been usually associated with an idealist current of neo-romantic theater. This apparent contradiction may be attributed to the author's Krausist ideas. In fact, Krausism is in essence a philosophical approach rooted in German idealism, that is, supported on liberal principles of freedom and democracy. In their study on the writer, Carmen Menéndez Onrubia and Julián Ávila Arellano assert that one of the most peculiar traits of Echegaray's dramatic work is the fusion of scientific pragmatism—"as a trained engineer, he composed drama using laboratory formulas"—and the regenerationist idealism of Krausism. See Carmen Menéndez Onrubia and Julián Ávila Arellano, *El neorromanticismo español y su época: Epistolario de José Echegaray a María Guerrero* (Madrid: Consejo Superior de Investigaciones Científicas, 1987), 16. This syncretism can also be seen in his political proposals, wherein the influence of his rigorous training as an engineer is evident in the way he approaches the country's problems. For a detailed account on Echegaray's facet as engineer, see Fornieles Alcaraz, *Trayectoria*, 31–59.

9. Echegaray, "Discurso Real Academia de Ciencias," 98.

10. These ideas resonate with the principles of national and national-state consolidation suggested by historian Ernest Gellner, according to which it was the demands of industrialization that led to the consolidation of a particular sense of collectivity. In this view, scientific knowledge and technological development constituted important parts of national pride—two forces that in conjunction with the joint efforts of society could generate the necessary conditions for modernization. See Ernest Gellner, *Nations and Nationalism* (Ithaca: Cornell University Press, 1983), 38ff.

11. Echegaray, "Discurso Real Academia de Ciencias," 100.

12. José Echegaray, *Discurso leído el día 10 de noviembre de 1898 en el Ateneo Científico, Literario y Artístico de Madrid* (Madrid: Establecimiento tipográfico Sucesores de Rivadeneyra, 1898), 28.

13. José Echegaray, "Recuerdos," *La España Moderna* 19, no. 223 (1907): 73.

14. José Echegaray, "Intervención de José Echegaray en las Cortes Constituyentes (Diario de Sesiones de las Cortes Constituyentes. Sesión del miércoles 5 de mayo de 1869: La cuestión religiosa)," *Arbor* 179, no. 707/708 (2004): 727.

15. José Echegaray, "Aplicación de las fuerzas naturales á la industria y al comercio," in *Conferencias del curso de 1879 á 1880. Círculo de la Unión Mercantil* (Madrid: Imprenta y estereotipia de El Liberal, 1881), 112.

16. Echegaray, "Aplicación de las fuerzas," 106.

17. Echegaray, "Aplicación de las fuerzas," 117.

18. Echegaray, "Aplicación de las fuerzas," 120 (brackets in original).

19. As a student of Gabriel Rodríguez, professor of political economy in the Escuela de Ingenieros de Caminos, Echegaray was introduced to the principles of free trade. Rodríguez had created the journal *El Economista* (The economist), a publication dedicated to analyzing economic matters from a liberal perspective. In reference to the principles defended by the journal's editorial line, Javier Fornieles highlights a clear preoccupation with wellness and improving living conditions, the belief that progress may be achieved by successfully exploiting national natural resources, and the valuing of science and technology as indispensable tools. Rodríguez also exposed Echegaray to the ideas of Frédéric Bastiat, a renowned economic philosopher who understood free trade as a flawless system that could reverse economic decline. See Fornieles Alcaraz, *Trayectoria*, 92.

20. As noted in chapter 1, one of the most influential theorists of Materialism was Ludwig Büchner. Widely discussed in intellectual circles, in Spain his famous 1855 study *Kraft und Stoff* had a profound impact on how industrialization was assimilated. Comparing the human being to a steam machine, the text contends that there is no such a thing as a spiritual energy or soul (understood here in Descartes's terms). As a consequence, there are no *spiritual forces* that differ in essence from any other material mechanisms: "The circulation of the blood is clearly mechanical, and the anatomical apparatus by which it is effected perfectly resembles that made by the hand of man. The heart has its valves, like a steam engine, and their closure produces audible sounds; the friction of the air against the walls of the *bronchiæ* produces the sound in ordinary breathing; the ascent of the blood from the inferior extremities to the heart is accomplished by mechanical arrangements. . . . It is also known that organic processes cannot be called self-acting, as they, like the changes in the inorganic world, are only called into action by the external world and the physical forces connected therewith. Hence, as Schaller observes, physiology begins now to reject the theory of an essential difference between the organic and inorganic world." See Ludwig Büchner, *Force and Matter: Empirico-Philosophical Studies, Intelligibly Rendered*, trans. J. Frederick Collingwood (London: Trübner, 1864), 220.

21. Echegaray, "Discurso Real Academia de Ciencias," 102.

22. José Echegaray, "Recuerdos," *La España Moderna* 16, no. 190 (1904): 73.

23. José Echegaray, "Discurso leído por José Echegaray ante la Real Academia Española en su recepción pública el día 20 de mayo de 1894," *Arbor* 179, no. 707/708 (2004): 737.

24. Within the political instability that Fernando VII's demise produced in the country, *common legality* was the name of the legal frame that guaranteed the coexistence of opposite political ideologies. As a progressive proposal, this formulation intended to stimulate progress through the discussion of different perspectives. The philosophical problems and political implications of implementing common legality in Spain were widely studied by thinkers such as Emilio Castelar, Francesc Cambó, Manuel Pando Fernández de Pinedo, Marqués de Miraflores, and Juan López Serrano. See, for example, Juan López Serrano, *La legalidad común: Solución política* (Madrid: Imprenta y litografía de Nicolás González, 1875), 30; and Ignacio Fernández Sarasola, *Los partidos políticos en el pensamiento español: De la Ilustración a nuestros días* (Madrid: Marcial Pons, 2009), 167–176.

25. Echegaray, "Discurso Real Academia Española," 743.

26. Echegaray, "Discurso Real Academia Española," 749–750.

27. Echegaray, "Discurso Real Academia Española," 782.

28. See Francesc Artal, "El pensamiento económico de Pere Estasen i Cortada," *Revista de Historia Económica* 7, no. 1 (1989): 75.

29. For a detailed examination of the ways in which doctrines of protectionism and free trade were projected as beneficial and essential programs to boost the industrial economy, see Antón Costas Comesaña, "El librecambio, la industrialización y sus desencantos: Argumentos a favor del proteccionismo en la segunda mitad del siglo XIX," in *Industrialización en España: Entusiasmos, desencantos y rechazos. Ensayos en homenaje al profesor Fabián Estapé*, ed. Germà Bel i Queralt and Alejandro Estruch Manjón (Madrid: Civitas, 1997), 205–226.

30. See Anson Rabinbach, *The Human Motor: Energy, Fatigue, and the Origins of Modernity* (New York: BasicBooks, 1990), 71.

31. Pere Estasen i Cortada, *La protección y el libre cambio: Consideraciones generales sobre la organización económica de las nacionalidades y la libertad de comercio* (Barcelona: Establecimiento tipográfico de los Sucesores de Ramírez, 1880), 23.

32. Estasen i Cortada, *La protección*, 25.

33. Estasen i Cortada, *La protección*, 31–32.

34. Estasen i Cortada, *La protección*, 86.

35. Estasen i Cortada, *La protección*, 89.

36. Estasen i Cortada, *La protección*, 87.

37. For Estasen, the United States's success in its expansion and modernization was exemplary; he paid particular attention to how the country allocated local capital in material and social projects such as the railroad expansion or the creation of educational institutions: "[En Estados Unidos,] a la sombra de la protección sábiamente aplicada, que produce la verdadera competencia, la acumulación de capitales dió impulso extraordinario á la iniciativa particular, la cual compitió en los ferro-carriles, que abarataron extraordinariamente sus precios de transporte, facilitando el tráfico interior; en las industrias particulares, en los artefactos mas sencillos, en los servicios de la vida pública y privada, y hasta, pudiéramos decir, en la administración" ([In the United States,] under a wisely applied protectionist system that redounded to true competition, capital accumulation gave extraordinary impulse to private initiatives, which competed in projects such as the railroad development, advancement that produced an extraordinary reduction in transportation costs, making internal trade easier. Private initiatives also participated in independent industries, the creation of the simplest devices, private and public services, and even, we can say, in the state administration). See Estasen i Cortada, *La protección*, 91.

38. Estasen i Cortada, *La protección*, 93.

39. Joshua Goode, *Impurity of Blood: Defining Race in Spain, 1870–1930* (Baton Rouge: Louisiana State University Press, 2009), 78ff.

40. Estasen i Cortada, *La protección*, 94.

41. Estasen i Cortada, *La protección*, 299.

42. Manuel Julivert Casagualda, *Una historia de la geología en España: En su contexto socioeconómico, cultural y político, y en el marco de la geología internacional* (Barcelona: Publicacions i Edicions de la Universitat de Barcelona, 2014), 132.

43. Steven L. Driever, "'And Since Heaven Has Filled Spain with Goods and Gifts': Lucas Mallada, the Regenerationist Movement, and the Spanish Environment, 1881–90," *Journal of Historical Geography* 24, no. 1 (1998): 47.

44. According to cultural critic Pedro Ribas, regenarationists' assessment of the country evolved in different, distinctive stages: initially, and from an objective perspective generally supported by evidence collection and data analysis, authors like Mallada, Ángel Ganivet, Joaquín Costa, Ricardo Macías Picavea, or Damián Isern anticipated the decline of the country, even before the loss of its imperial status in 1898. After the Spanish-American War, regenerationism moved toward a more philosophical interpretation of national problems. In this second period, the work of writers such as Miguel de Unamuno, Azorín, and Antonio Machado, all associated with the Generación del 98, gained special attention. Finally, the line of thinking proposed by these authors was adopted and

extended during the first two decades of the twentieth century by writers like Ramiro de Maeztu and José Ortega y Gasset, many of whose ideas were used to articulate the government ideology of Antonio Maura (1921–1922) and later the dictatorship of Miguel Primo de Rivera (1923–1930). See Pedro Ribas, "Regeneracionismo: una relectura," in *El Regeneracionismo en España: Política, educación, ciencia y sociedad*, ed. Vicent Salavert Fabiani and Manuel Suárez Cortina (Valencia: Universitat de València, 2007), 47–80.

45. In the words of Driever, "Mallada's lucid, hard-hitting prose and his ability to relate science to everyday life suited Costa's plans to change the *Boletín* from an outlet exclusively for Institución Libre professors to a lively vehicle for the most renowned publicists of [Spain]." See Driever, "Lucas Mallada," 41.

46. See Driever, "Lucas Mallada, 43–45.

47. Gellner, *Nations and Nationalism*, 72.

48. See José Luis Ramos Gorostiza, "'Carácter nacional' y decadencia en el pensamiento español," *Biblio 3W. Revista Bibliográfica de Geografía y Ciencias Sociales* 15, no. 860 (2010), http://www.ub.edu/geocrit/b3w-860.htm (accessed February 15, 2021).

49. Regenerationism reformulated the idea of the national character that emerged during the Romantic era. By the beginning of the nineteenth century, philosophical idealism strongly influenced European thought. Thinkers like Johann Herder were pivotal in the development of notions of nation and nationality. Herder considered that each nation had a unique collective consciousness, a particular spirit that he denominated *Volksgeist*. These spiritual traits were connected with transhistorical ethnic, cultural, psychological, and geographic traits. Such traits could also be passed on from parents to their offspring, conforming to what Laura Otis refers to as "organic memory." See Laura Otis, *Organic Memory: History and the Body in the Late Nineteenth and Early Twentieth Centuries* (Lincoln: University of Nebraska Press, 1994), 94–95. From this perspective, a nation was considered an organic entity that could not only be sick but also treated. Important regenerationists such as Mallada, Costa, Macías Picavea, Luis Morote, Damián Isern, Rafael Altamira, Miguel de Unamuno, and Ángel Ganivet attempted a complete identification of the problematic traits that caused this illness. See, for example, Joseph Harrison, "The Regenerationist Movement in Spain after the Disaster of 1898," *European History Quarterly* 9, no. 1 (1979): 7–8.

50. Lucas Mallada, *Los males de la patria y la futura revolución española: Consideraciones generales acerca de sus causas y efectos*, ed. Francisco J. Flores Arroyuelo (Madrid: Alianza, 1969), 15.

51. Mallada, *Los males*, 18.

52. Mallada, *Los males*, 31–32.

53. In the context of the nineteenth-century consolidation of national identities, the work of Friedrich Ratzel (1844–1904) was essential to integrating geography and culture. Influenced by Herder's notion of the "national spirit," and the materialist principles embedded in Darwin's and Haeckel's biological evolutionism, Ratzel's work departed from the idea that natural and environmental conditions determine the social-group behavior. This concept challenged topographic notions that saw geography as a fixed entity, thus problematizing connections between ideology and science. Ratzel's ideas also had a profound impact on Spanish regenerationism. See Driever, "Lucas Mallada," 42; and Santos Casado de Otaola, *Naturaleza patria: Ciencia y sentimiento de la naturaleza en la España del regeneracionismo* (Madrid: Marcial Pons, 2010), 96.

54. Mallada, *Los males*, 38.

55. Mallada, *Los males*, 20.
56. Mallada, *Los males*, 40.
57. Mallada, *Los males*, 54.
58. Lyell's pioneering contributions to the emerging discipline of geology in the early 1800s set the foundations for understanding the slow and uniform changes in Earth's topography. He noted that rates of erosion, for instance, were uniform and practically imperceptible; as a consequence, it seemed as if landscapes' attributes were immutable. See Julivert, *Una historia*, 73–77.
59. Mallada, *Los males*, 43.
60. Mallada, *Los males*, 49.
61. Mallada, *Los males*, 138–139.
62. Mallada, *Los males*, 149.
63. Mallada, *Los males*, 41.
64. Lucas Mallada, "Los progresos de la Geología en España durante el siglo XIX," in *Discursos leídos ante la Real Academia de Ciencias Exactas, Físicas y Naturales en la recepción pública del Sr. D. Lucas Mallada y Pueyo el día 29 de junio de 1897* (Madrid: Imprenta de Luis Aguado, 1897), 66.
65. Mallada, "Los progresos de la Geología," 12.
66. Mallada, "Los progresos de la Geología," 63–64.
67. Lucas Mallada, *Proyecto de una nueva división territorial de España* (Madrid: Imprenta y estereotipia de El Liberal, 1881), 10.

CHAPTER 3 — THE EDUCATIONAL ENGINE

1. Studies have shown that by 1900, 70 percent of the population in the country was illiterate, and only a very small percentage of the rural population and urban lower classes had access to secondary instruction. See, for example, José María Ortiz Villajos, *Tecnología y desarrollo económico en la historia contemporánea: Estudio de las patentes registradas en España entre 1882 y 1935* (Madrid: Oficina Española de Patentes, 1999), 7ff.
2. José Echegaray, "Discurso leído por José Echegaray ante la Real Academia de Ciencias Exactas, Físicas y Naturales en su recepción pública el 11 de marzo de 1866," *Revista de Obras Públicas*, no. 14 (1866): 101.
3. See José Manuel Sánchez Ron, *Cincel, martillo y piedra: Historia de la ciencia en España (siglos XIX y XX)* (Madrid: Taurus, 2000), 332ff.
4. See Ricardo Macías Picavea, *El problema nacional: Hechos, causas, remedios* (Madrid: Librería general de Victoriano Suárez, 1899), 157.
5. Gumersindo Vicuña, *Discurso leído en la Universidad Central en el acto de apertura del curso académico de 1875 a 1876* (Madrid: Imprenta de José M. Ducazcal, 1875), 33. In the original, "segunda enseñanza" refers to the courses that follow elementary school, which in the United States are divided into middle school and high school. Currently, universities in Spain require a selection test and specific scores for admission in their programs.
6. Vicuña, *Discurso Universidad Central*, 55 (emphasis in original).
7. Vicuña, *Discurso Universidad Central*, 57.
8. Vicuña, *Discurso Universidad Central*, 69.
9. Vicuña, *Discurso Universidad Central*, 77.
10. Laureano Calderón, "Discurso leído el 1 de febrero de 1884 en la inauguración de los debates de la sección de ciencias exactas físicas y naturales," in *Discursos leídos en el*

Ateneo Científico, Literario y Artístico de Madrid con motivo de la apertura del curso de 1884 (Madrid: Imprenta Central, 1884), 103–104.

11. See Ángela del Valle, *Aportación bio-bibliográfica a la historia de la ciencia: Universidad Central 1886–1902* (Madrid: Narcea, 1998), 109–110.

12. Calderón, "Discurso leído el 1 de febrero," 103.

13. Laureano Calderón, *Discurso leído en la Universidad Central en la inauguración del curso académico de 1892 a 1893* (Madrid: Imprenta Colonial, 1892), 7–8.

14. Calderón, "Discurso leído el 1 de febrero," 132 (my emphasis).

15. See Calderón, "Discurso leído el 1 de febrero," 101.

16. Calderón, "Discurso leído el 1 de febrero," 107.

17. During the late nineteenth century, physicists tried to unify under the same discipline disparate observable phenomena, including heat, electricity, magnetism, and light. This unity was achieved through the conceptualization of energy. As Barry Gold suggests, "Energy Physics had an undeniable appeal for those who sought the one law to rule them all (one law to bind them), who sought the *first principles* from which all others could be derived." See Barry Gold, *ThermoPoetics: Energy in Victorian Literature and Science* (Cambridge: MIT, 2010), 72 (emphasis in original).

18. Calderón, "Discurso leído el 1 de febrero," 116.

19. Calderón, *Discurso Universidad Central*, 50.

20. Calderón, *Discurso Universidad Central*, 35.

21. See Jo Labanyi, *Gender and Modernization in the Spanish Realist Novel* (Oxford: Oxford University Press, 2000), 49.

22. Calderón, "Discurso leído el 1 de febrero," 131.

23. Although Krausist liberals such as Calderón believed in the need to improve the well-being of the working classes, they also validated Herbert Spencer's idea that difficult conditions were a necessary price for preserving the natural order of the social apparatus. For a detailed discussion of the Krausist articulation of *La cuestión social* (the social question), see Labanyi, *Gender and Modernization*, 52–64.

24. Calderón, *Discurso Universidad Central*, 50.

25. Calderón, "Discurso leído el 1 de febrero," 134.

26. Calderón, *Discurso Universidad Central*, 76.

27. Zoel García de Galdeano, "La matemática y su enseñanza," pt. 4, *El Progreso Matemático (Segunda Serie)* 2, no. 8 (1900): 53.

28. As noted in previous chapters, by the end of the nineteenth century the country's decadence was considered a disease derived, among other causes, from the absence of scientific, technological, and industrial development. See, for example, Agustí Nieto-Galán, "A Republican Natural History in Spain around 1900: Odón de Buen (1863–1945) and His Audiences," *Historical Studies in the Natural Sciences* 42, no. 3 (2012): 164; and Leoncio López-Ocón, *Breve historia de la ciencia Española* (Madrid: Alianza, 2003), 305, 334ff.

29. The assessments of José Echegaray, Santiago Ramón y Cajal, and, to some extent, the ideas of García de Galdeano's disciple Julio Rey Pastor are also relevant in this regard. See, for example, Echegaray's and Ramón y Cajal's inauguration speeches in the Real Academia de Ciencias Exactas, Físicas y Naturales in 1866 and 1897, respectively, and Rey Pastor's inauguration speech for the Asociación Española para el Progreso de las Ciencias (Spanish Association for the Progress of Science) in 1915. As for the relationship between scientific development and the consolidation of the modern nation-state, see, among

others, Jorge Cañizares-Esguerra, *Nature, Empire, and Nation: Explorations of the History of Science in the Iberian World* (Palo Alto: Stanford University Press, 2006), 23, 41; Ernest Gellner, *Nations and Nationalism* (Ithaca: Cornell University Press, 1983), 118ff; and Agustí Nieto-Galán, "The Images of Science in Modern Spain: Rethinking the 'Polémica,'" in *The Sciences in the European Periphery during the Enlightenment*, ed. Kostas Gavroglu (Dordrecht: Springer, 1999), 77ff.

30. See José Sala Catalá, *Ideología y ciencia biológica en España entre 1860 y 1881. La difusión de un paradigma* (Madrid: Consejo Superior de Investigaciones Científicas, 1987), 43.

31. Issue 48 of the journal featured a detailed list of subscribers that counted at least eighty entries, including names like José Echegaray, Eduardo Torroja, the Ateneo de Madrid, the Real Academia de Ciencias de Madrid, and the Facultad de Ciencias de Barcelona. See Zoel García de Galdeano, "Lista de los señores suscriptores en 1894," *El Progreso Matemático* 4, no. 48 (1894): 371–372.

32. Authors such as Rafael Altamira, Joaquín Costa, and Manuel de la Revilla were among the scientists, intellectuals, and politicians who condemned this dependency but, paradoxically, had been educated outside Spain. See Jean Antoine Díaz, "La ciencia española: La paradoja romántica del discurso regeneracionista," in *Pensamiento y literatura en España en el siglo XIX: Idealismo, positivismo, espiritualismo*, ed. Yves Lissorgues and Gonzalo Sobejano (Toulouse: Presses Universitaires du Mirail, 1998), 295–310.

33. Some advances in mathematics made during the first half of the nineteenth century challenged the methodological preoccupation of realism, and in particular naturalism, with the communication of truth through literature. Naturalist writers addressed problems by classifying the actions, determining their causes, explaining the effects; they had to operate following a set of principles linking the environment and the individual. In that sense, mathematics served as model, theory, and metaphor for society. Science, and especially scientific ideas and developments, functioned as a web of interactions wherein language and culture could be articulated around the notion of progress. See N. Katherine Hayles, *The Cosmic Web: Scientific Field Models and Literary Strategies in the Twentieth Century* (Ithaca: Cornell University Press, 1984), 9–11, 21.

34. See Andrea Henderson, "Math for Math's Sake: Non-Euclidean Geometry, Aestheticism, and *Flatland*," *PMLA* 124, no. 2 (2009): 457.

35. García de Galdeano, "La matemática y su enseñanza," pt. 1, *El Progreso Matemático (Segunda Serie)* 1, no. 2 (1899): 39.

36. García de Galdeano, "La matemática," 40.

37. Zoel García de Galdeano, "A nuestros lectores," *El Progreso Matemático* 2, no. 13 (1892): 6.

38. Due to financial problems, *El Progreso* closed its doors in 1895, an unfortunate outcome that evidenced once again the precarious situation of scientific promotion in the country. This lack of support, from both official and private sources, highlights the quixotic nature of García de Galdeano's endeavor. In response to García Alix's reforms, the journal briefly reappeared for two more years, beginning in 1899. See Mariano Hormigón Blánquez, "*El Progreso Matemático* (1891–1900): Un estudio sobre la primera revista matemática española," *Llull: Revista de la Sociedad Española de Historia de las Ciencias y de las Técnicas* 4, no. 6–7 (1981): 88.

39. Hormigón Blánquez, "El Progreso," 88.

40. Zoel García de Galdeano, "La enseñanza de la ciencia matemática en la Universidad," *El Progreso Matemático* 4, no. 38 (1894): 86 (my emphasis).

41. Studies on the history of the press have noted the strong connections between politics and print media. See, for example, Jean-François Botrel, *Libros, prensa y lectura en la España del siglo XIX* (Madrid: Fundación Germán Sánchez Ruipérez, 1993); and María Sánchez Agustí, *La educación española a finales del XIX: una mirada a través del periódico republicano* La Libertad (Lleida: Milenio, 2002).

42. At the moment *El Progreso* initiated its activities in 1891, projects of educational reform had been limited to those of the Plan Pidal (1845), the Plan Pastor Díaz (1847), and the Plan Seijas (1850). Some of these early attempts to reformulate higher education had ended up doing more harm than good. The Plan Seijas, for instance, centralized the university system authority in Madrid, making it impossible for other institutions in the system to confer the degree of doctor in sciences. In 1854 circumstances changed with the creation of the Ministerio de Fomento. Under its support, in 1857 the Plan Moyano made one of the most important contributions to modernizing education by integrating the teaching of science into the different projects of national development. See Mariano Peset and José Luis Peset, "Las universidades españolas del siglo XIX y las ciencias," *Ayer*, no. 7 (1992): 30–34.

43. Zoel García de Galdeano, "Estudios sobre la enseñanza y el organismo de la ciencia matemática," pt. 1, *El Progreso Matemático* 3, no. 25 (1893): 10.

44. García de Galdeano, "La enseñanza," 67.

45. Zoel García de Galdeano, "Resumen de algunas consideraciones expuestas por el Señor Menéndez Pelayo," *El Progreso Matemático* 4, no. 38 (1894): 57.

46. Zoel García de Galdeano, "El objeto y los propósitos de la actual publicación." *El Progreso Matemático* 1, no. 1 (1891): 2.

47. García de Galdeano, "Estudios," 8.

48. See García de Galdeano, "Estudios," 8.

49. García de Galdeano, "Estudios," 15.

50. García de Galdeano, "La enseñanza," 95.

51. In addition to the works analyzed throughout this chapter, it is worth mentioning Santiago Ramón y Cajal's influential essay *Los tónicos de la voluntad* (1897), translated later into English as *Advice for a Young Investigator*. As he contends, a successful exploration of nature required more than just technical knowledge; it also implied a moral discipline whose cultivation would separate scientists from technicians. See Santiago Ramón y Cajal, *Los tónicos de la voluntad: reglas y consejos sobre investigación científica* (Buenos Aires: Espasa-Calpe, 1971), 25–32.

52. See "José Rodríguez Mourelo," Real Academia de la Historia, http://dbe.rah.es/biografias/25713/jose-rodriguez-mourelo (accessed February 17, 2021); and Stefan Pohl-Valero, "The 'Circulation' of Energy: Thermodynamics, National Culture, and Social Progress in Spain, 1868–1890," in *Popularizing Science and Technology in the European Periphery, 1800–2000*, ed. Faidra Papanelopoulou, Agustí Nieto-Galán, and Enrique Perdiguero (Burlington: Ashgate, 2009), 131.

53. José Rodríguez Mourelo, "Discurso del Ilmo. Sr. D. José Rodríguez Mourelo," in *Discursos leídos ante la Real Academia de Ciencias Exactas, Físicas y Naturales en la recepción pública del Ilmo. Sr. D. José Rodríguez Mourelo el 24 de mayo de 1903* (Madrid: Imprenta de Luis Aguado, 1903), 16.

54. Rodríguez Mourelo, "Discurso del Ilmo. Sr. D. José Rodríguez Mourelo," 20.

55. Rodríguez Mourelo, "Discurso del Ilmo. Sr. D. José Rodríguez Mourelo," 32.

56. José Echegaray, "Discurso del Excmo. Sr. D. José Echegaray," in *Discursos leídos ante la Real Academia de Ciencias Exactas, Físicas y Naturales en la recepción pública del Ilmo. Sr. D. José Rodríguez Mourelo el 24 de mayo de 1903* (Madrid: Imprenta de Luis Aguado, 1903), 49.

57. Echegaray, "Discurso del Excmo. Sr. D. José Echegaray," 51.

58. Herder's notion of the existence of a national genius—collective features that define a country's psychology—was particularly embraced by Francisco Giner de los Ríos, one of Sanz del Río's followers and most prominent promoters of Krausist ideas on education, for whom literature and philosophy were the measures of the affective and intellectual breadth of national culture. See Juan López-Morillas, *El krausismo español: Perfil de una aventura intelectual* (México: Fondo de Cultura Económica, 1980), 87–89, 115; and Solomon Lipp, *Francisco Giner de los Ríos: A Spanish Socrates* (Waterloo: Wilfrid Laurier University Press, 1985), 49–53.

59. See Alfredo Sosa-Velasco, *Médicos escritores en España 1885–1955* (Woodbridge: Tamesis, 2010), 3.

CHAPTER 4 — SOCIAL ENGINEERING

1. See Ramiro Reig, "Vicente Blasco Ibáñez (1867–1928): Promotor de rebeldías," in *Liberales, agitadores y conspiradores. Biografías heterodoxas del siglo XIX*, ed. Isabel Burdiel and Manuel Pérez Ledesma (Madrid: Espasa, 2000), 331–361; and *Vicente Blasco Ibáñez* (Madrid: Espasa, 2002).

2. See Gloria Elsa Li, "Literatura y sociedad en *El intruso* y *La bodega* de Vicente Blasco Ibáñez" (PhD diss., Stanford University, 1995), 18–32.

3. When referring to Blasco's most famous novel, *Los cuatro jinetes del Apocalipsis* (*The Four Horsemen of the Apocalypse*), Paul Smith relates how the Valencian writer never thought that his work would bring him fame and wealth at the levels it did once the novel was translated into English in 1918, calling the attention of film producers in Hollywood. Neither could he imagine that this work would make him the most famous Spanish person in the United States. See Paul Smith, *Vicente Blasco Ibáñez: Una nueva introducción a su vida y obra* (Santiago de Chile: Andrés Bello, 1972), 36.

4. Among this group, one can find the regionalist novels: *Arroz y Tartana* (Airs and graces) (1894), *La barraca* (*The Cabin*) (1898) and *Cañas y barro* (*Reeds and Mud*) (1902). Likewise, critics have focused on the study of novels that had recognition due to their historical or cultural import: for example, *Sangre y arena* (*Blood and Sand*) (1908), *Los cuatro jinetes del Apocalipsis* (1916), or *Mare Nostrum* (Our sea) (1918). See, among others, Pablo Torres, "Político, periodista, escritor . . . Blasco Ibáñez, un personaje incómodo," *Carta de España*, no. 737 (2017): 32–33; Carlos Blanco Aguinaga, *Juventud del 98* (Barcelona: Crítica, 1978); and *De restauración a restauración: Ensayos sobre literatura, historia e ideología* (Sevilla: Renacimiento, 2007); Roland Forgues, "El universo narrativo de las novelas sociales de Vicente Blasco Ibáñez," *Letras de Deusto* 8, no. 15 (1978): 69–157; and *Vicente Blasco Ibáñez: Mito y realidad* (Barcelona: Puvill, 1987); José Luis León Roca, *Vicente Blasco Ibáñez* (Valencia: Ajuntament de Valencia, 1997); and Smith, *Blasco Ibáñez*.

5. See, for example, Manuel Muñoz Cortés, "Variedad regional, lengua vernácula y conflicto lingüístico en el Bilbao del siglo XIX y su función en *El intruso* de Blasco Ibáñez." In *Estudis en memòria del professor Manuel Sanchis Guarner* (Valencia: Universitat de València, 1984); José María de Areilza, "Cuatro libros sobre Bilbao," *Revista de Occidente*,

no. 50 (1985): 65–82; Li, "Literatura y sociedad"; and Rosa Eugenia Montes Doncel, "Variaciones de un tema novelesco: *Doña Perfecta* de Galdós y *El intruso* de Blasco Ibáñez," *Revista Hispánica Moderna* 58, no. 1–2 (2005): 39–59.

6. Vicente Blasco Ibáñez, *El intruso* (Valencia: F. Sempere y Cia., 1904), 414.

7. Blasco Ibáñez, *El intruso*, 56–57.

8. The rapid expansion of the iron mining industry and its subsidiaries illustrates the proportion and pace of this region's transformation. Historian Rafael Uriarte explains this process, highlighting how Biscay slowly transitioned from disorganized and archaic practices of mineral exploitation to a technological and mechanized industry using sophisticated methods of extraction and processing. See "La minería vizcaína del hierro en las primeras etapas de la industrialización," in *La industrialización del norte de España*, ed. Jordi Maluquer de Motes, Emiliano Fernández de Pinedo, and José Luis Hernández Marco (Barcelona: Crítica, 1988), 157.

9. Blasco Ibáñez, *El intruso*, 58.

10. Blasco Ibáñez, *El intruso*, 26.

11. Blasco Ibáñez, *El intruso*, 194.

12. Blasco Ibáñez, *El intruso*, 194–195 (my emphasis).

13. By the end of nineteenth century, the transactional economy had replaced what Leigh Mercer identifies as "signifiers of bourgeois affluence and social power," confirming the bourgeoisie's eagerness to emulate the aristocracy by becoming financial actors in the national economy. See Leigh Mercer, *Urbanism and Urbanity: The Spanish Bourgeois Novel and Contemporary Customs (1845–1925)* (Lewisburg: Bucknell University Press, 2013), 107.

14. Blasco's idea of progress was aligned with socialist principles that had had a positive reception in Spain at the turn of the century. The writer shared the vision of many politically involved groups of workers who saw in socialism an answer to their claims: better living conditions and "the subordination of individual desires to humanitarian universal principles." See Radoslav Andrea Tsanoff, *Civilization and Progress* (Lexington: University Press of Kentucky, 1971), 121.

15. See Miguel Artola, *Antiguo Régimen y revolución liberal* (Barcelona: Ariel, 1979), 8.

16. Blasco Ibáñez, *El intruso*, 72.

17. Blasco Ibáñez, *El intruso*, 303.

18. Blasco Ibáñez, *El intruso*, 324 (my emphasis).

19. Blasco Ibáñez, *El intruso*, 157–158.

20. Blasco Ibáñez, *El intruso*, 235.

21. See, for example, Miguel Barrios, *Plan de estudios para los niños* (Madrid: Tipografía de Manuel Ginés Hernández, 1888), 13ff; Orencio Garcés Banzo, *Programa de industria para uso de las escuelas de niños* (Logroño: Imprenta y librería de El Riojano, 1878), 15ff; or Julián López Catalán, *La industria y el comercio al alcance de los niños* (Barcelona: J. Bastinos e Hijo, 1874), 5ff.

22. See Forgues, *Blasco: Mito y realidad*, 16.

23. Blasco Ibáñez, *El intruso*, 118.

24. The Church showed great interest in the labor question, a concern that was confirmed in the promulgation of the 1891 encyclical *Rerum Novarum*. This document, written by Pope Leo XIII, reevaluated the role of the Church as a conciliatory agent in the tension between the classes in power and the proletariat, reaffirming at the same time Catholicism's social nature. According to the ideological guidelines proposed by the text,

workers had rights over the product of their labor, an empowerment that would ameliorate the sense of alienation and expropriation that characterized industrial production. This declaration of principles, however, was only the Church's attempt to regain space in the reorganization of powers of the industrial economy. In reality, the Church's role as an unconditional ally of aristocracy and industrial bourgeoisie had not changed. Even though some sectors of the Church felt a stronger commitment to the labor cause, in general its representatives continued ignoring the social unbalance and extreme conditions that industrialization had imposed on workers. See Feliciano Montero García, *El primer catolicismo social y la Rerum Novarum (1899–1902)* (Madrid: Consejo Superior de Investigaciones Científicas, 1983), 95–100.

25. See Juan Pablo Fusi, *Política obrera en el país vasco: 1880–1923* (Madrid: Turner, 1975), 229–230.

26. Ramiro de Maeztu, "El socialismo bilbaíno," *Germinal*, Julio 16, 1897.

27. Ramiro de Maeztu, "La obra del odio," *El Heraldo de Madrid*, Octubre 28, 190[illegible].

28. Inspired by similar imagery, Marx explained: "The worker puts his life into the object; but now his life no longer belongs to him but to the object . . . the more the worker spends himself, the more powerful becomes the alien world of objects which he creates over and against himself." See Karl Marx, "Estranged Labor," in *Princeton Readings in Political Thought: Essential Texts Since Plato*, ed. Mitchell Cohen and Nicole Fermon (Princeton: Princeton University Press, 1996), 440.

29. M. K. Flynn, *Ideology, Mobilization, and the Nation: The Rise of Irish, Basque, and Carlist Nationalist Movements in the Nineteenth and Early Twentieth Centuries* (New York: Macmillan, 2000), 120–121.

30. Blasco Ibáñez, *El intruso*, 116.

31. Blasco Ibáñez, *El intruso*, 309.

32. See Fusi, *Política obrera*, 199.

33. The reception of *Los vencedores* was severely affected by the wrong classification that publishers gave it, which prevented audiences from properly understanding the work's social criticism and denunciatory character. *Los vencidos*, instead, was received as a necessary denunciation that exposed with great detail and objectivism workers' multiple struggles in the mining sector, thus becoming a valuable historical document for understanding the social tensions of this period. See Cecilio Alonso, "Vida y obra de Manuel Ciges Aparicio" (PhD diss., Universidad Complutense de Madrid, 1985), 2:202–205.

34. See José Álvarez Junco, "El minero como creación literaria," in *Mineros, sindicalismo y política*, ed. José Ángel Fernández Villa (Madrid: Fundación José Barreiro, 1987), 435.

35. See Víctor Fuentes, "La literatura comprometida de Miguel Ciges Aparicio," *Ínsula: Revista de Letras y Ciencias Humanas* 27, no. 305 (1972): 13.

36. Álvarez Junco, "El minero," 438.

37. Manuel Ciges Aparicio, *Las luchas de nuestros días: Los vencidos* (Madrid: Librería de los Sucesores de Hernando, 1910), 65.

38. Ciges Aparicio, *Los vencidos*, 72.

39. Ciges Aparicio, *Los vencidos*, 75.

40. Ciges Aparicio, *Los vencidos*, 99.

41. Ciges Aparicio, *Los vencidos*, 102.

42. Ciges Aparicio, *Los vencidos*, 158.

43. Manuel Ciges Aparicio, *Las luchas de nuestros días: Los vencedores* (Madrid: M. Pérez Villavicencio, 1908), 43.
44. Ciges Aparicio, *Los vencedores*, 83.
45. Ciges Aparicio, *Los vencedores*, 144.
46. Ciges Aparicio, *Los vencedores*, 147.
47. Ciges Aparicio, *Los vencedores*, 202.
48. Ciges Aparicio, *Los vencedores*, 215.
49. Elizabeth Rojas-Auda, "Concha Espina y su obra comprometida: 1888–1936" (PhD diss., University of Wisconsin–Madison, 1997), 85.
50. Concha Espina, *El metal de los muertos* (Madrid: Renacimiento, 1920), 22 (my emphasis).
51. Espina, *El metal*, 27.
52. Espina, *El metal*, 43.
53. Espina, *El metal*, 67.
54. Espina, *El metal*, 106.
55. Espina, *El metal*, 106.
56. Espina, *El metal*, 115.
57. Espina, *El metal*, 189.
58. Espina, *El metal*, 121.
59. Espina, *El metal*, 119.
60. Espina, *El metal*, 220.
61. Rojas-Auda, "Concha Espina," 103.

CHAPTER 5 — TECHNOLOGIES OF MASS DIFFUSION

1. Darwinism had a profound cultural impact on Spain. Yet, given that few people could really approach the complexity of Darwin's work, the debate became a discussion on the incompatibility between science and religion. See Dale J. Pratt, *Signs of Science: Literature, Science, and Spanish Modernity since 1868* (West Lafayette: Purdue University Press, 2001), and "'Fundamental Truths' to 'Playthings of Science': Science, Technology, and Modernity in Pardo Bazán," in *Approaches to Teaching the Writings of Emilia Pardo Bazán*, ed. Margot Versteeg and Susan Walter (New York: Modern Language Association of America, 2017), 71–77.
2. See Thomas F. Glick, ed., *The Comparative Reception of Darwinism* (Chicago: University of Chicago Press, 1988), 336.
3. See, for example, José Echegaray, "Discurso del excmo señor D. José Echegaray," in *Discursos leídos ante La Real Academía Española en la recepción pública del excmo señor D. Francisco Silvela* (Madrid: Establecimiento tipográfico Sucesores de Rivadeneyra, 1893), 59.
4. José Rodríguez Carracido, *Discurso leído en la solemne primera adjudicación de la medalla de su nombre al Excmo. Sr. D. José Echegaray el día 16 de junio de 1907* (Madrid: Imprenta de La Gaceta de Madrid, 1907), 9.
5. José Echegaray, *Discursos leídos ante la Real Academia de Ciencias Exactas, Fisicas y Naturales en la recepción pública del sr. D. Francisco de Paula Rojas* (Madrid: Imprenta de Luis Aguado, 1894), 5.
6. See, Echegaray, *Discursos recepción Francisco Rojas*, 33.
7. José Echegaray, *Teorías modernas de la física: Unidad de las fuerzas materiales*, vol. 1 (Madrid: Imprenta y estereotipia de M. Rivadeneyra, 1873), 58.

8. Echegaray, *Teorías modernas*, 1:55 (emphasis in original).

9. Jose Echegaray, *Teorías modernas de la física: Unidad de las fuerzas materiales*, vol. 2 (Madrid: Imprenta y librería de J. Gaspar, 1883), 174–175 (emphasis in original).

10. Echegaray, *Teorías modernas*, 2:196.

11. Echegaray, *Teorías modernas*, 2:205.

12. Echegaray, *Teorías modernas*, 2:207–208.

13. Echegaray, *Teorías modernas*, 2:210.

14. Echegaray, *Teorías modernas*, 2:212.

15. A prolific journalist, in recent years specialists have turned their attention to Pardo Bazán's work in multiple capacities (correspondent, reporter, critic, opinion writer) and for a significant number of publications with a wide and diverse readership. See, among others, José Manuel González Herrán and Cristina Patiño Eirín, *Emilia Pardo Bazán: El periodismo* (A Coruña: Real Academia Galega, 2007); González Herrán, "Reescritura en algunas crónicas periodísticas de Emilia Pardo Bazán (1912–1915)," in *Escritoras españolas en los medios de prensa: 1868–1936*, ed. Carmen Servén and Ivana Rota (Sevilla: Renacimiento, 2014), 117–137; Pilar Palomo Vázquez, Pilar Vega Rodríguez, and Concepción Núñez Rey, eds., *Emilia Pardo Bazán, Periodista* (Madrid: Arco Libros, 2015); and Emilia Pérez Romero, *El periodismo de Emilia Pardo Bazán* (Vigo: Editorial Academia del Hispanismo, 2016).

16. Emilia Pardo Bazán, *La cuestión palpitante* (Madrid: Imprenta Central, 1883), 131.

17. Among the many proposals that entered this debate, essays such as *El positivismo materialista* (Materialist positivism) (1872) by Zeferino González, "El doctor Büchner o el catecismo de los materialistas" (Dr. Büchner or the materialists' catechism) (1873) by Francisco Caminero, *Examen del materialismo moderno* (Examination of modern materialism) (1875) by Antonio María Fable, or "El materialismo contemporáneo y la ciencia cristiana" (Contemporary materialism and Christian science) (1877) by Luis María Eleizalde, gained wide recognition. See Stefan Pohl-Valero, "The 'Circulation' of Energy: Thermodynamics, National Culture, and Social Progress in Spain, 1868–1890," in *Popularizing Science and Technology in the European Periphery, 1800–2000*, ed. Faidra Papanelopoulou, Agustí Nieto-Galán, and Enrique Perdiguero (Burlington: Ashgate, 2009), 121.

18. The discovery of the two first laws of thermodynamics produced an ideological division within the scientific community in Europe. This difference of opinions resulted from the diverse interpretations that those laws had in a theological context. On the one hand, the principle of conservation of energy, or the first law, coincided with the theory of evolution proposed by Darwin in problematizing the idea of a static and finite cosmos. This principle suggested a universe in constant transformation, one that was ruled by natural laws and where the idea of God had no place. As noted earlier, Herbert Spencer in England and Ludwig Büchner in Germany, among other thinkers, promoted these notions and appropriated them as social principles. On the other hand, entropy, or the second law, suited religious precepts by supporting the idea of a universe in a progressive and terminal degeneration. Following Kant's vitalist principles, in this vision God was the igniter of all energy in the universe, and nature, as the Old Testament had prophesied, was then in charge of taking it to its end. See Favio Cala Vitery and Stefan Pohl-Valero, "Energía, entropía y religión. Un repaso histórico," *Revista de la Academia Colombiana de Ciencias Exactas, Físicas y Naturales* 34, no. 130 (2010): 46–47.

19. Laura Otis, "Science and Signification in the Early Writings of Emilia Pardo Bazán," *Revista de Estudios Hispánicos* 29, no. 1 (1995): 82.

20. Emilia Pardo Bazán, "La ciencia amena," pt. 1, *Revista Compostelana* 1, no. 3 (1876): 18.

21. Pardo Bazán, "La ciencia amena," pt. 8, *Revista Compostelana* 1, no. 11 (1876): 83.

22. Pardo Bazán, "La ciencia amena," pt. 11, *Revista Compostelana* 2, no. 14 (1876): 107.

23. Though less common, misinterpretations of *The Origin* also produced the obfuscation of the notion of sexual selection. In spite of the fact that the first translations into Spanish of *The Origin* and of *The Descent of Man* (1871), either from French or directly from English, were not necessarily inaccurate or incomplete, discussions of Darwin's ideas in Spain were conditioned more by political ideologies or religious prejudices than by the actual scientific content presented in the texts. See Travis Landry, *Subversive Seduction: Darwin, Sexual Selection, and the Spanish Novel* (Seattle: University of Washington Press, 2012), 3–23, 72–101.

24. Emilia Pardo Bazán, "Reflexiones científicas contra el darwinismo," pt. 4. *La Ciencia Cristiana*, no. 5 (1877): 492.

25. Pardo Bazán, "Reflexiones," pt. 4. *La Ciencia Cristiana*, no. 5 (1877): 482.

26. Pardo Bazán, "Reflexiones," pt. 1, *La Ciencia Cristiana*, no. 4 (1877): 291.

27. Emilio Huelin, "Petroleo cual agente de bárbaros incendios y de otras distintas maneras considerado," *Revista de España* 31, no. 131 (1873): 9.

28. Emilia Pardo Bazán, "Sud-Exprés," in *Obras completas*, vol. 36 (Madrid: Pueyo, 1900), 5.

29. It is worth noting here that Pardo Bazán's story is only one of many examples of literature inspired by the problems of socialization that new transportation technologies posed, or by the particular ways in which travel conditions stimulated imagination and encouraged observation. In 1853, for instance, Pedro Antonio de Alarcón published his short novel *El clavo* (*The Nail*), in which the conversation that the narrator initiates with an unknown woman while traveling in a stagecoach from Granada to Malaga leads to solving a crime. See Alarcón, *El clavo* (Madrid: Rodríguez Serra, 1900). As I briefly discussed in chapter 1, within this same tradition, and inscribed in Madrid's urban realm, the short novel *La novela en el tranvía* (1871) by Benito Pérez Galdós uses the vehicle's space for introspection as a multilayer narrative framework in which a drama of passion and betrayal defies the limits between reality and imagination. See Pérez Galdós, "La novela en el tranvía," in *13 cuentos*, ed. Esteban Gutiérrez Díaz-Bernardo (Madrid: Edaf, 2001), 171–201.

30. Pardo Bazán, "Sud-Exprés," 10.

31. Pardo Bazán, "Sud-Exprés," 6.

32. Pardo Bazán, "Sud-Exprés," 8.

33. The allusion to the cinematographer in Pardo Bazán's story is significant. "Sud-Exprés" was published in 1902, only six years after Georges Méliès's projection of the film *Arrivée d'un train (Gare de Vincennes)* (Train's arrival at Vincennes's station), a piece that subverted the notions of fiction and reality by producing in the spectator the illusion of being in front of a moving train. Cinematograph's technical advances produced a complete reorganization of perception, changing, as in Pardo Bazán's story, the rules that defined the appropriation of reality. It is probable that Pardo Bazán saw the film in one of her many visits to Paris. See Catharina Vallejo, "Emilia Pardo Bazán, Gender, Modernity and Nationalism at the Paris World Exhibitions of 1889 and 1900," *Revista Canadiense de Estudios Hispánicos* 32, no. 3 (2008): 454, 465–466.

34. Pardo Bazán, "Sud-Exprés," 5.

35. See María Isabel Jiménez Morales, "Emilia Pardo Bazán, cronista en París (1889)," *Revista de Literatura* 52, no. 140 (2008): 523.

36. Emilia Pardo Bazán, "La vida contemporánea," *La Ilustración Artística* 18, no. 929 (Octubre 16, 1899): 666.

37. Pardo Bazán, "La vida contemporánea." *La Ilustración Artística* 20, no. 1031 (Septiembre 30, 1901): 634.

38. Even when the initiative to establish common time zones around the world came from the popularization of the telegraph, it was only with the expansion of railroad networks that projects of normalization reached the legislators, becoming a national priority in countries like England or France: "Despite all the good scientific and military arguments for world time, it was the railroad companies and not the governments that were the first to institute it." See Stephen Kern, *The Culture of Time and Space, 1880–1918* (Cambridge: Harvard University Press, 2003), 12.

39. Pardo Bazán, "La vida contemporánea," *La Ilustración Artística* 19, no. 958 (Mayo 7, 1900): 298.

40. Pardo Bazán, "La vida contemporánea." *La Ilustración Artística*, 33, no. 1697 (Julio 6, 1914): 446.

41. Pardo Bazán, "La vida contemporánea, 446.

42. Pardo Bazán, "La vida contemporánea," 634.

43. William W. Stowe, *Going Abroad: European Travel in Nineteenth-Century American Culture* (Princeton: Princeton University Press, 1994), 11.

44. See Michel de Certeau, *The Practice of Everyday Life*, trans. Steven Rendall (Berkeley: University of California Press, 2002), 111–115.

45. Georg Simmel, *Simmel on Culture: Selected Writings*, ed. David Frisby and Mike Featherstone (London: Sage, 2000), 141.

46. Quoted in Lily Litvak, "Abolición del tiempo y el espacio. El tren a fines del siglo XIX," *Compás de Letras. Monografías de Literatura Española*, no. 7 (1995): 248.

47. Alarcón, Pedro Antonio de, *Viajes por España* (Madrid: Imprenta de A. Pérez Durrul, 1883), 7.

48. See, among others, Fernando García Lara, ed., *En torno a Pedro Antonio de Alarcón* (Granada: Diputación Provincial de Granada, 1993); and Antonio Lara Ramos, *Pedro Antonio de Alarcón* (Granada: Comares, 2001).

49. See Cristina Viñes Millet, "El africanismo de Pedro Antonio de Alarcón," in *Pedro Antonio de Alarcón y la guerra de África: Del entusiasmo romántico a la compulsión colonial*, ed. José Antonio González Alcantud and Manuel Lorente Rivas (Barcelona: Anthropos, 2004), 46.

50. Alarcón, *Viajes*, 311 (my emphasis).

51. Alarcón, *Viajes*, 271–272.

52. Alarcón, *Viajes*, 22.

53. Alarcón, *Viajes*, 23.

54. See Alarcón, *Viajes*, 25.

55. Alarcón, *Viajes*, 40–41.

56. Alarcón, *Viajes*, 77 (my emphasis).

57. Alarcón, *Viajes*, 81.

58. Alarcón, *Viajes*, 97–98.

59. Alarcón, *Viajes*, 85.

60. Alarcón, *Viajes*, 284 (emphasis in original).

61. Alarcón, *Viajes*, 287.

62. Alarcón, *Viajes*, 277.

CHAPTER 6 — INDUSTRIAL FOOTPRINT

1. See Timothy Clark, *The Value of Ecocriticism* (Cambridge: Cambridge University Press, 2019), 111.

2. Clark, *The Value*, 114.

3. Emilia Pardo Bazán, *Obras completas*, vol. 19, *Al pie de la Torre Eiffel* (Madrid: Establecimiento tipográfico de Idamor Moreno, 1899), 197–199.

4. See Peter J. Taylor, *Political Geography: World-Economy, Nation-State and Locality* (London: Longman, 1985), 177–179.

5. See, among others, Carlos Blanco Aguinaga, *La historia y el texto literario: tres novelas de Galdós* (Madrid: Nuestra Cultura, 1978); William Hutchinson Shoemaker, *The Novelistic Art of Galdós*, 3 vols. (Valencia: Albatros Hispanófila, 1980–1982); Stephen Gilman, *Galdós and the Art of the European Novel, 1867–1887* (Princeton: Princeton University Press, 1981); Peter A. Bly, *Galdos's Novel of the Historical Imagination: A Study of the Contemporary Novels* (Liverpool: Francis Cairns, 1983); Brian J. Dendle, *Galdós: The Early Historical Novels* (Columbia: University of Missouri Press, 1986); Akiko Tsuchiya, *Images of the Sign: Semiotic Consciousness in the Novels of Benito Pérez Galdós* (Columbia: University of Missouri Press, 1990); Ignacio-Javier López, *Galdós y el arte de la prosa* (Barcelona: Promociones y Publicaciones Universitarias, 1993); and Geoffrey Ribbans, *History and Fiction in Galdós's Narratives* (Oxford: Clarendon, 1993).

6. See María Paz Yáñez, "El dilema discursivo en *Marianela*," *Anales Galdosianos*, no. 29–30 (1994–1995): 51–62; Joaquín Casalduero, "El tren como símbolo: El progreso, la clase social, la cibernética en Galdós," *Anales Galdosianos*, no. 5 (1970): 15–22; and C. A. Jones, "Galdós's *Marianela* and the Approach to Reality," *Modern Language Review* 56, no. 4 (1961): 515–519.

7. See, for example, Peter A. Bly, "Egotism and Charity in *Marianela*," *Anales Galdosianos*, no. 7 (1972): 49–64; Andrew A. Anderson, "Necessary Sacrifices: From Romanticism to Naturalism in Galdós's *Marianela*," *Bulletin of Spanish Studies* 92, no. 6 (2015): 914ff; and Sarah Sierra, "Time and the Environment: 'Slow Violence' in Benito Pérez Galdós's *Doña Perfecta* and *Marianela*," *Confluencia: Revista Hispánica de Cultura y Literatura* 35, no. 1 (2019): 48ff.

8. See Juan López-Morillas, *El krausismo español: Perfil de una aventura intelectual* (México: Fondo de Cultura Económica, 1980), 136.

9. Anderson, "Necessary Sacrifices," 912.

10. Jones, "Galdós's *Marianela*," 515. Comte's evolutionary model of social change includes a notion of development that goes from simple to complex forms of interaction. Comte's aim in studying society in these terms was to understand the principles of social order. See Robert C. Scharff, *Comte after Positivism* (Cambridge: Cambridge University Press, 2002), 73–75.

11. These profound and irreversible changes in what Timothy Clark has framed as the "Earth system" can be understood under the critical lens of ecocriticism. From this perspective, the kind of transformation of nature for the benefit of human progress that we can see in Galdós's novel, for example, forms part of a newly recognized context often termed *Anthropocene*. For a critical analysis of the cultural implication of these concepts, see Clark, *The Value*, 11, 18–28.

12. Benito Pérez Galdós, *Marianela* (Madrid: Imprenta y litografía de La Guirnalda, 1878), 15–16.

13. Pérez Galdós, *Marianela*, 18.

14. Pérez Galdós, *Marianela*, 30.

15. Pérez Galdós, *Marianela*, 59.

16. Pérez Galdós, *Marianela*, 61.

17. Pérez Galdós, *Marianela*, 278.

18. Jones, "Galdós's *Marianela*," 519.

19. Transposing this dichotomy between idealization and reality onto the literary realm, Anderson contends that *Marianela* "is not solely a novel about science, industry and progress, about mining and ophthalmology, and their respective benefits and drawbacks, but also, simultaneously, a novel about analogous changes occurring in the world of literature and in particular narrative, namely the transition in literary modes from Romanticism to Naturalism." See Anderson, "Necessary Sacrifices," 921.

20. The novel can be read from the perspective of platonic idealism. Galdós builds the basic argument following certain metaphors found in Plato's argumentation: contrasts between human form and essence, body and soul, articulated through binaries like matter and substance or light and obscurity, thus illustrate the complex configuration of fin-de-siglo society. See Mario E. Ruiz, "El idealismo platónico en *Marianela* de Galdós," *Hispania: A Journal Devoted to the Teaching of Spanish and Portuguese* 53, no. 4 (1970): 871.

21. Pérez Galdós, *Marianela*, 139.

22. Pérez Galdós, *Marianela*, 262.

23. Pérez Galdós, *Marianela*, 143.

24. See, respectively, Pérez Galdós, *Marianela*, 66, 67, 141, 218.

25. Sierra, "Time and the Environment," 53.

26. Pérez Galdós, *Marianela*, 50.

27. Jones, "Galdós's *Marianela*," 519.

28. See María Pilar Aparici Llanas, *Las novelas de tesis de Benito Pérez Galdós* (Barcelona: Consejo Superior de Investigaciones Científicas, 1982), 13–15.

29. León Bodevin, "José Ortega Munilla y la novela del siglo XIX," (PhD diss., Texas Tech University, 1999), 34.

30. Since by the mid-nineteenth century, Spain did not count with the necessary infrastructure to communicate its urban centers to the country's periphery, the railroad consolidated itself as an answer to national fragmentation. See Pilar Lozano Carbayo, *El libro del tren: 150 años de ferrocarril en España* (Madrid: Guillomía Comunicacion Gráfica, 1988); and Antonio Gómez Mendoza, *Ferrocarril, industria y mercado en la modernización de España* (Madrid: Espasa Calpe, 1989), 201ff.

31. Ortega Munilla's novel *Don Juan Solo* (1880) played an important role in conceptualizing the two parts of *La desheredada* (*The Disinherited Lady*), one of Galdós's most influential novels, that appeared the same year. See Ignacio-Javier López, "Ortega Munilla y la doble génesis de *La desheredada*," *Anales Galdosianos* 20, no. 2 (1985): 7–17; and Michael A. Schnepf, "A Note on Galdós, Ortega Munilla, and *La desheredada*," *Romance Notes* 39, no. 1 (1998): 3–7.

32. See Ruth Schmidt, *Ortega Munilla y sus novelas* (Madrid: Revista de Occidente, 1973); and León Bodevin and Jorge Medina, "La novela burguesa de Ortega Munilla: Paradigma del despotismo ideológico decimonónico," *Horizontes: Revista de la Universidad Católica de Puerto Rico* 46, no. 90 (2004): 81–101.

33. José Ortega Munilla, *El tren directo, relación contemporánea* (Madrid: Oficinas de La Ilustración Española y Americana, 1880), 109.

34. During the second half of the nineteenth century, schools of engineering developed a particular personality in students and future technicians that set their mission and role in the society of the future. It is not surprising then, as León Bodevin explains, that in Ortega Munilla's novel it is precisely an engineer who has the important mission of understanding nature by exercising critical judgment and using the power of reason. See Bodevin, "José Ortega Munilla," 61.

35. Ortega Munilla, *El tren*, 117.

36. The case of José María de Salamanca i Mayol, first Marquis of Salamanca, is typical of the practices depicted in the novel. A pioneer of modernization, the reputation of this aristocrat was obscured by his role in the 1846–1848 economic crisis. Accused of speculation and embezzlement, the marquis abused his position as secretary of finances (Ministro de Hacienda) to benefit his personal businesses. This misuse of power affected the whole economy of the country and created distrust of the convenience and importance of modernization. See Miguel Ángel López-Morell, "Salamanca y la construcción del ferrocarril de Aranjuez," in *Ferrocarril y Madrid: Historia de un progreso*, ed. Francisco Polo Muriel, María Jesús Matilla Quiza, and Manuel Benegas Capote (Madrid: Fundación de los Ferrocarriles Españoles, 2002), 13–44.

37. Ortega Munilla, *El tren*, 55–56.

38. Ortega Munilla, *El tren*, 241–242.

39. Ortega Munilla, *El tren*, 34.

40. Ortega Munilla, *El tren*, 35.

41. Ortega Munilla, *El tren*, 243.

42. Ortega Munilla, *El tren*, 266.

43. Ortega Munilla, *El tren*, 243.

44. Ortega Munilla, *El tren*, 36.

45. Ortega Munilla, *El tren*, 244.

46. See Leopoldo Alas, *Solos de Clarín* (Madrid: Alianza, 1971), 289.

47. Ortega Munilla, *El tren*, 82–83.

48. Ortega Munilla, *El tren*, 56.

49. Kelly Elizabeth Sultzbach, *Ecocriticism in the Modernist Imagination: Forster, Woolf, and Auden* (Cambridge: Cambridge University Press, 2016), 13.

50. Palacio Valdés regarded *La aldea perdida* as one of his best works as it was a delicate portrait of his childhood's lost paradise. See Brian J. Dendle, *Spain's Forgotten Novelist: Armando Palacio Valdés, 1853–1938* (Lewisburg: Bucknell University Press, 1995), 106–107.

51. Dendle, *Spain's Forgotten Novelist*, 81.

52. Armando Palacio Valdés, *La espuma* (Barcelona: Imprenta de Henrich y Compañía en comandita, 1890), 179–180.

53. Palacio Valdés, *La espuma*, 199–200.

54. See Guillermo Gortázar, *Alfonso XIII, hombre de negocios: persistencia del Antiguo Régimen, modernización económica y crisis política, 1902–1931* (Madrid: Alianza, 1986), 217.

55. Palacio Valdés, *La espuma*, 201–202.

56. See Dendle, *Spain's Forgotten Novelist*, 17.

57. In *The Division of Labor in Society* (1893), Emile Durkheim hypothesizes that what held preindustrial societies together were values, sentiments, and norms, equally shared

by all. Modern forms of social organizations, instead, distorted these collective codes of conduct and solidarity, making differences between traditional ways of living and industrial labor unmistakable. This disruption to the ways in which labor, but also community engagement, was allocated led to the idealization of the past and the demonization of the present. See Lewis A. Coser, introduction to *The Division of Labor in Society*, by Emile Durkheim, trans. W. D. Halls (New York: Free Press, 1984), xiv.

58. Armando Palacio Valdés, *La aldea perdida: Novela-poema de costumbres campesinas* (Madrid: Imprenta de los Hijos de M. G. Hernández, 1903), 2.

59. Palacio Valdés, *La aldea*, 276.

60. Guadalupe Gómez-Ferrer Morant, *Palacio Valdés y el mundo social de la Restauración* (Oviedo: Instituto de Estudios Asturianos, 1983), 403.

61. Palacio Valdés, *La aldea*, 127.

62. Palacio Valdés, *La aldea*, 79.

63. Palacio Valdés, *La aldea*, 86.

64. Palacio Valdés, *La aldea*, 84–85.

65. Palacio Valdés, *La aldea*, 108.

66. Palacio Valdés, *La aldea*, 65.

67. Palacio Valdés, *La aldea*, 238–239.

68. Palacio Valdés, *La aldea*, 39.

69. Palacio Valdés, *La aldea*, 228.

70. Palacio Valdés, *La aldea*, 155.

71. Discussing the violent and abrupt conclusion of the novel, critics have agreed that there is a profound pessimism in Palacio Valdés's vision of the future. Dendle, for example, explains the end as the collision of two temporalities: "The brutal slayings of Jacinto and Demetria symbolize . . . the destruction of the Asturian rural Arcadia by the invasion of industrial civilization, with its corresponding disruption of an age-old pastoral and agricultural society." See Dendle, *Spain's Forgotten Novelist*, 108.

72. Palacio Valdés, *La aldea*, 242–243.

Bibliography

Alarcón, Pedro Antonio de. *El clavo*. Madrid: Rodríguez Serra, 1900.

———. *De Madrid a Nápoles*. Madrid: Imprenta y librería de Gaspar y Roig, 1861.

———. *Diario de un testigo de la Guerra de África*. Madrid: Imprenta y librería de Gaspar y Roig, 1859.

———. *La Alpujarra: Sesenta leguas a caballo precedidas de seis en diligencia*. Madrid: Imprenta y librería de Miguel Guijarro, 1874.

———. *Viajes por España*. Madrid: Imprenta de A. Pérez Durrul, 1883.

Alas, Leopoldo. *Solos de Clarín*. Madrid: Alianza, 1971.

Alonso, Cecilio. "Vida y obra de Manuel Ciges Aparicio." 3 vols. PhD diss., Universidad Complutense de Madrid, 1985. Biblioteca Nacional de España (5/55306).

Alonso, Gregorio. "How to Be Religious under Liberalism." In *Spain in the Nineteenth Century: New Essays on Experiences of Culture and Society*, edited by Andrew Ginger and Geraldine Lawless, 89–108. Manchester: Manchester University Press, 2018.

Álvarez Junco, José. "El minero como creación literaria." In *Mineros, sindicalismo y política*, edited by José Ángel Fernández Villa, 429–439. Madrid: Fundación José Barreiro, 1987.

———. *Mater dolorosa: La idea de España en el siglo XIX*. Madrid: Taurus, 2003.

Anderson, Andrew A. "Necessary Sacrifices: From Romanticism to Naturalism in Galdós's *Marianela*." *Bulletin of Spanish Studies* 92, no. 6 (2015): 907–929.

Aparici Llanas, María Pilar. *Las novelas de tesis de Benito Pérez Galdós*. Barcelona: Consejo Superior de Investigaciones Científicas, 1982.

Appadurai, Arjun. *The Social Life of Things: Commodities in Cultural Perspective*. Cambridge: Cambridge University Press, 2001.

Areilza, José María de. "Cuatro libros sobre Bilbao." *Revista de Occidente*, no. 50 (1985): 65–82.

Aronna, Michael. *"Pueblos enfermos": The Discourse of Illness in the Turn-of-the-Century Spanish and Latin American Essay*. Chapel Hill: University of North Carolina Press, 1999.

Artal, Francesc. "El pensamiento económico de Pere Estasen i Cortada." *Revista de Historia Económica* 7, no. 1 (1989): 71–78.

Artola, Miguel. *Antiguo Régimen y revolución liberal*. Barcelona: Ariel, 1979.

Barrios, Miguel. *Plan de estudios para los niños*. Madrid: Tipografía de Manuel Ginés Hernández, 1888.

Beaumont, Matthew, and Michael J. Freeman. *The Railway and Modernity: Time, Space, and the Machine Ensemble*. Oxford: Peter Lang, 2007.

Beer, Gillian. *Open Fields: Science in Cultural Encounter*. Oxford: Oxford University Press, 1995.

Bel i Queralt, Germà, and Alejandro Estruch Manjón, eds. *Industrialización en España: Entusiasmos, desencantos y rechazos. Ensayos en homenaje al profesor Fabián Estapé*. Madrid: Civitas, 1997.

Benjamin, Walter. "The Work of Art in the Age of Its Technological Reproducibility." In *The Work of Art in the Age of Its Technological Reproducibility, and Other Writings on Media*, edited by Brigid Doherty, Michael W. Jennings, and Thomas Y. Levin, 19–55. Cambridge: Belknap Press of Harvard University Press, 2008.

Bernal, J. D. *Science and Industry in the Nineteenth Century*. London: Routledge, 2011.

Blanco Aguinaga, Carlos. *De restauración a restauración: Ensayos sobre literatura, historia e ideología*. Sevilla: Renacimiento, 2007.

———. *Juventud del 98*. Barcelona: Crítica, 1978.

———. *La historia y el texto literario: Tres novelas de Galdós*. Madrid: Nuestra Cultura, 1978.

Blasco Ibáñez, Vicente. *El intruso*. Valencia: F. Sempere y Cia., 1904.

———. *La bodega*. Valencia: F. Sempere y Cia., 1905.

———. *La catedral*. Valencia: F. Sempere y Cia., 1903.

———. *La horda*. Valencia: F. Sempere y Cia., 1905.

Bly, Peter A. "Egotism and Charity in *Marianela*." *Anales Galdosianos*, no. 7 (1972): 49–64.

———. *Galdos's Novel of the Historical Imagination: A Study of the Contemporary Novels*. Liverpool: Francis Cairns, 1983.

Bodevin, León. "José Ortega Munilla y la novela del siglo XIX." PhD diss., Texas Tech University, 1999. ProQuest (304527062).

Bodevin, León, and Jorge Medina. "La novela burguesa de Ortega Munilla: Paradigma del despotismo ideológico decimonónico." *Horizontes: Revista de la Universidad Católica de Puerto Rico* 46, no. 90 (2004): 81–101.

Boivin, Nicole. *Material Cultures, Material Minds: The Impact of Things on Human Thought, Society, and Evolution*. Cambridge: Cambridge University Press, 2008.

Borbón, Juan Carlos I de. "Palabras de Su Majestad el Rey en la inauguración de la nueva Línea de Alta Velocidad Madrid-Valencia." *Casa de S.M. el Rey* (Diciembre 18, 2010). https://www.casareal.es/EN/Actividades/Paginas/actividades_discursos_detalle.aspx?data=2482 (accessed February 1, 2021).

Botella, Federico de. "De cómo nuestro suelo no es tan pobre como se quiere decir." *Boletín de la Institución Libre de Enseñanza*, no. 6 (1882): 137–140, 161–163.

Botrel, Jean-François. *Libros, prensa y lectura en la España del siglo XIX*. Madrid: Fundación Germán Sánchez Ruipérez, 1993.

Bourdieu, Pierre. *The Field of Cultural Production: Essays on Art and Literature*. Edited and introduced by Randal Johnson. Cambridge: Polity, 1999.

Broder, Albert, Miguel Á. Pérez de Perceval Verde, Alejandro Sánchez Rodríguez, and Carmen Marchán Sanz, eds. *La inversión extranjera en la minería española*. Madrid: Instituto Geológico y Minero de España, 2014.

Büchner, Ludwig. *Force and Matter: Empirico-Philosophical Studies, Intelligibly Rendered*. Translated by J. Frederick Collingwood. London: Trübner, 1864.

———. *Fuerza y materia: Estudios populares de historia y filosofía naturales*. Translated by A. Avilés. Madrid: Librería de Fernando Fe, 1878.

Burdiel, Isabel, and Manuel Pérez Ledesma, eds. *Liberales, agitadores y conspiradores. Biografías heterodoxas del siglo XIX*. Madrid: Espasa, 2000.

Cala Vitery, Favio, and Stefan Pohl-Valero. "Energía, entropía y religión. Un repaso histórico." *Revista de la Academia Colombiana de Ciencias Exactas, Físicas y Naturales* 34, no. 130 (2010): 37–52.

Calderón, Laureano. "Discurso leído el 1 de febrero de 1884 en la inauguración de los debates de la sección de ciencias exactas físicas y naturales." In *Discursos leídos en el Ateneo Científico, Literario y Artístico de Madrid con motivo de la apertura del curso de 1884*, 99–136. Madrid: Imprenta Central, 1884.

———. *Discurso leído en la Universidad Central en la inauguración del curso académico de 1892 a 1893*. Madrid: Imprenta Colonial, 1892.

———. "La cuestión social y las ciencias naturales." In *Discurso pronunciado en el Ateneo de Madrid en 1891*, 1–11. Madrid: Moya, 1891.

Candelaria, Michael. *The Revolt of Unreason: Miguel de Unamuno and Antonio Caso on the Crisis of Modernity*. Amsterdam: Rodopi, 2012.

Cano Ballesta, Juan. *Literatura y tecnología: Las letras españolas ante la Revolución Industrial, 1890–1940*. Valencia: Pre-Textos, 1999.

Cánovas del Castillo, Antonio. *Discurso pronunciado el día 6 de noviembre de 1882 en el Ateneo Científico y Literario de Madrid con motivo de la apertura de sus cátedras*. Madrid: Imprenta Central a cargo de Víctor Saiz, 1882.

Cañizares-Esguerra, Jorge. "Iberian Science in the Renaissance: Ignored How Much Longer?" *Perspectives on Science* 12, no. 1 (2004): 86–124.

———. *Nature, Empire, and Nation: Explorations of the History of Science in the Iberian World*. Palo Alto: Stanford University Press, 2006.

Capellán de Miguel, Gonzalo. *De Orovio a Cossío: Vieja y nueva educación, La Rioja (1833–1933)*. Logroño: Instituto de Estudios Riojanos, 1999.

———. "Política educativa bajo los gobiernos de Cánovas y Sagasta: Propuestas para una interpretación." *Berceo. Revista Riojana de Ciencias Sociales y Humanidades*, no. 139 (2000): 123–144.

Carr, Raymond et al., eds. *Imágenes y ensayos del 98*. Valencia: Fundación Cañada Blanch, 1998.

Casado de Otaola, Santos. *Naturaleza patria: Ciencia y sentimiento de la naturaleza en la España del regeneracionismo*. Madrid: Marcial Pons, 2010.

Casalduero, Joaquín. "El tren como símbolo: El progreso, la clase social, la cibernética en Galdós." *Anales Galdosianos*, no. 5 (1970): 15–22.

Caudet, Francisco. *El parto de la modernidad: la novela española de los siglos XIX y XX*. Madrid: De la Torre, 2002.

Cayuela Fernández, José Gregorio. *Un siglo de España: Centenario 1898–1998*. Ciudad Real: Universidad de Castilla La Mancha, 1998.

Certeau, Michel de. *The Practice of Everyday Life*. Translated by Steven Rendall. Berkeley: University of California Press, 2002.

Ciges Aparicio, Manuel. *Las luchas de nuestros días: Los vencedores*. Madrid: M. Pérez Villavicencio, 1908.

———. *Las luchas de nuestros días: Los vencidos*. Madrid: Librería de los Sucesores de Hernando, 1910.

Clark, Timothy. *The Value of Ecocriticism*. Cambridge: Cambridge University Press, 2019.

Clarke, Bruce. *Energy Forms: Allegory and Science in the Era of Classical Thermodynamics*. Ann Arbor: University of Michigan Press, 2001.

Cohen, Mitchell, and Nicole Fermon, eds. *Princeton Readings in Political Thought: Essential Texts Since Plato*. Princeton: Princeton University Press, 1996.

Coser, Lewis A. Introduction to *The Division of Labor in Society*, by Emile Durkheim, translated by W. D. Halls, ix–xxiv. New York: Free Press, 1984.

Costas Comesaña, Antón. "El librecambio, la industrialización y sus desencantos: Argumentos a favor del proteccionismo en la segunda mitad del siglo XIX." In *Industrialización en España: Entusiasmos, desencantos y rechazos. Ensayos en homenaje al profesor Fabián Estapé*, edited by Germà Bel i Queralt and Alejandro Estruch Manjón, 205–226. Madrid: Civitas, 1997.

Cruz, Jesús. *The Rise of Middle-Class Culture in Nineteenth-Century Spain*. Baton Rouge: Louisiana State University Press, 2011.

Daly, Nicholas. *Literature, Technology, and Modernity, 1860–2000*. Cambridge: Cambridge University Press, 2004.

Davis, Ryan A., and Alicia Cerezo Paredes, eds. *Modernity and Epistemology in Nineteenth-Century Spain: Fringe Discourses*. Lanham: Lexington Books, 2017.

Del Río Joan, Francisco. "La electricidad en España al advenimiento de Don Alfonso XIII." In *La ciencia y la industria eléctrica en España al subir al trono S.M. el Rey Don Alfonso XIII*. Madrid: La Energía Eléctrica, 1902.

Deleuze, Gilles. *Difference and Repetition*. Translated by Paul Patton. New York: Columbia University Press, 1994.

Delmiro Coto, Benigno. *Literatura y minas en la España de los siglos XIX y XX*. Gijón: Trea, 2003.

Dendle, Brian J. *Galdós: The Early Historical Novels*. Columbia: University of Missouri Press, 1986.

———. *Spain's Forgotten Novelist: Armando Palacio Valdés, 1853–1938*. Lewisburg: Bucknell University Press, 1995.

Díaz, Jean Antoine. "La ciencia española: la paradoja romántica del discurso regeneracionista." In *Pensamiento y literatura en España en el siglo XIX: Idealismo, positivismo, espiritualismo*, edited by Yves Lissorgues and Gonzalo Sobejano, 295–310. Toulouse: Presses Universitaires du Mirail, 1998.

Doherty, Brigid, Michael W. Jennings, and Thomas Y. Levin, eds. *The Work of Art in the Age of Its Technological Reproducibility, and Other Writings on Media*. Cambridge: Belknap Press of Harvard University Press, 2008.

Driever, Steven L. "'And Since Heaven Has Filled Spain with Goods and Gifts': Lucas Mallada, the Regenerationist Movement, and the Spanish Environment, 1881–90." *Journal of Historical Geography* 24, no. 1 (1998): 36–52.

Echegaray, José. "Aplicación de las fuerzas naturales á la industria y al comercio." In *Conferencias del curso de 1879 á 1880. Círculo de la Unión Mercantil*, 105–125. Madrid: Imprenta y estereotipia de El Liberal, 1881.

———. "Discurso del excmo. Sr. D. José Echegaray." In *Discursos leídos ante la Real Academia de Ciencias Exactas, Físicas y Naturales en la recepción pública del Ilmo.*

Sr. D. José Rodríguez Mourelo el 24 de mayo de 1903, 45–59. Madrid: Imprenta de Luis Aguado, 1903.

———. "Discurso del excmo. señor D. José Echegaray." In *Discursos leídos ante La Real Academia Española en la recepción pública del excmo señor D. Francisco Silvela*, 43–53. Madrid: Establecimiento tipográfico Sucesores de Rivadeneyra, 1893.

———. *Discurso leído el día 10 de noviembre de 1898 en el Ateneo Científico, Literario y Artístico de Madrid*. Madrid: Establecimiento tipográfico Sucesores de Rivadeneyra, 1898.

———. "Discurso leído por José Echegaray ante la Real Academia de Ciencias Exactas, Físicas y Naturales en su recepción pública el 11 de marzo de 1866." *Revista de Obras Públicas*, no. 14 (1866): 93–103.

———. "Discurso leído por José Echegaray ante la Real Academia Española en su recepción pública el día 20 de mayo de 1894." *Arbor* 179, no. 707/708 (2004): 731–784.

———. *Discursos leídos ante la Real Academia de Ciencias Exactas, Físicas y Naturales en la recepción pública del sr. D. Francisco de Paula Rojas*. Madrid: Imprenta de Luis Aguado, 1894.

———. "Intervención de José Echegaray en las Cortes Constituyentes (Diario de Sesiones de las Cortes Constituyentes. Sesión del miércoles 5 de mayo de 1869: La cuestión religiosa)." *Arbor* 179, no. 707/708 (2004): 715–729.

———. "Recuerdos." *La España Moderna* 16, no. 190 (1904): 56–76.

———. "Recuerdos." *La España Moderna* 19, no. 223 (1907): 68–79.

———. *Teorías modernas de la física: Unidad de las fuerzas materiales*. Vol. 1. Madrid: Imprenta y estereotipia de M. Rivadeneyra, 1873.

———. *Teorías modernas de la física: Unidad de las fuerzas materiales*. Vol. 2. Madrid: Imprenta y librería de J. Gaspar, 1883.

Espina, Concha. *El metal de los muertos*. Madrid: Renacimiento, 1920.

Estasen i Cortada, Pere. *La protección y el libre cambio: Consideraciones generales sobre la organización económica de las nacionalidades y la libertad de comercio*. Barcelona: Establecimiento tipográfico de los Sucesores de Ramírez, 1880.

Fernández de los Ríos, Ángel. *Guía de Madrid, manual del madrileño y del forastero*. Madrid: Oficinas de La Ilustración Española y Americana, 1876.

Fernández Sarasola, Ignacio. *Los partidos políticos en el pensamiento español: De la Ilustración a nuestros días*. Madrid: Marcial Pons, 2009.

Fernández Villa, José Ángel, ed. *Mineros, sindicalismo y política*. Madrid: Fundación José Barreiro, 1987.

Flynn, M. K. *Ideology, Mobilization, and the Nation: The Rise of Irish, Basque, and Carlist Nationalist Movements in the Nineteenth and Early Twentieth Centuries*. New York: Macmillan, 2000.

Forgues, Roland. "El universo narrativo de las novelas sociales de Vicente Blasco Ibáñez." *Letras de Deusto* 8, no. 15 (1978): 69–137.

———. *Vicente Blasco Ibáñez: Mito y realidad*. Barcelona: Puvill, 1987.

Fornieles Alcaraz, Javier. *Trayectoria de un intelectual de la Restauración: José Echegaray*. Almería: Cajalmería, 1989.

Foucault, Michel. *The Archaeology of Knowledge; The Discourse on Language*. Translated by Rupert Swyer. New York: Pantheon, 1972.

———. *Discipline and Punish: The Birth of the Prison*. Translated by Alan Sheridan. New York: Vintage Books, 2012.

Fuentes, Víctor. "La literatura comprometida de Miguel Ciges Aparicio." *Ínsula: Revista de Letras y Ciencias Humanas* 27, no. 305 (1972): 13.

Fusi, Juan Pablo. "El legado del 98." In *Imágenes y ensayos del 98*, edited by Raymond Carr et al., 289–302. Valencia: Fundación Cañada Blanch, 1998.

———. *España: La evolución de la identidad nacional*. Madrid: Temas de hoy, 2000.

———. *Política obrera en el país vasco: 1880–1923*. Madrid: Turner, 1975.

Ganivet, Ángel. *Cartas finlandesas*. Madrid: Librería general de Victoriano Suárez, 1905.

———. *Idearium español*. Granada: Tip. Lit. Vda. e Hijos de Sabatel, 1897.

Garcés y Banzo, Orencio. *Programa de industria para uso de las escuelas de niños*. Logroño: Imprenta y librería de El Riojano, 1878.

García Camarero, Ernesto. *La ciencia española entre la polémica y el exilio: Dentro de la evolución de la ciencia europea*. Madrid: Megustaescribir, 2016.

García Camarero, Ernesto, and Enrique García Camarero. *La polémica de la ciencia española*. Madrid: Alianza, 1970.

García de Galdeano, Zoel. "A nuestros lectores." *El Progreso Matemático* 2, no. 13 (1892): 3–6.

———. "El objeto y los propósitos de la actual publicación." *El Progreso Matemático* 1, no. 1 (1891): 1–3.

———. "Estudios sobre la enseñanza y el organismo de la ciencia matemática." 3 parts. *El Progreso Matemático* 3, no. 25 (1893): 5–17; 3, no. 29 (1893): 163–167; 4, no. 41 (1894): 66–68.

———. "La enseñanza de la ciencia matemática en la universidad." *El Progreso Matemático* 4, no. 38 (1894): 65–96.

———. "La matemática y su enseñanza." 4 parts. *El Progreso Matemático (Segunda Serie)* 1, no. 2 (1899): 38–44; 1, no. 3 (1899): 72–77; 2, no. 7 (1900): 23–27; 2, no. 8 (1900): 49–54.

———. "Lista de los señores suscriptores en 1894," *El Progreso Matemático* 4, no. 48 (1894): 371–372.

———. "Resumen de algunas consideraciones expuestas por el Señor Menéndez Pelayo." *El Progreso Matemático* 4, no. 38 (1894): 54–58.

García Lara, Fernando, ed. *En torno a Pedro Antonio de Alarcón*. Granada: Diputación Provincial de Granada, 1993.

Gavroglu, Kostas, ed. *The Sciences in the European Periphery during the Enlightenment*. Dordrecht: Springer, 1999.

Gellner, Ernest. *Nations and Nationalism*. Ithaca: Cornell University Press, 1983.

Gies, David Thatcher. *The Theatre in Nineteenth-Century Spain*. Cambridge: Cambridge University Press, 1994.

Gilman, Stephen. *Galdós and the Art of the European Novel, 1867–1887*. Princeton: Princeton University Press, 1981.

Ginger, Andrew, and Geraldine Lawless, eds. *Spain in the Nineteenth Century: New Essays on Experiences of Culture and Society*. Manchester: Manchester University Press, 2018.

Glick, Thomas F., ed. *The Comparative Reception of Darwinism*. Chicago: University of Chicago Press, 1988.

Gold, Barri. *ThermoPoetics: Energy in Victorian Literature and Science*. Cambridge: MIT, 2010.

Gómez Mendoza, Antonio. *Ferrocarril, industria y mercado en la modernización de España*. Madrid: Espasa Calpe, 1989.

Gómez-Ferrer Morant, Guadalupe. *Palacio Valdés y el mundo social de la Restauración*. Oviedo: Instituto de Estudios Asturianos, 1983.

González Alcantud, José Antonio, and Manuel Lorente Rivas, eds. *Pedro Antonio de Alarcón y la guerra de África: Del entusiasmo romántico a la compulsión colonial*. Barcelona: Anthropos, 2004.

González Cuevas, Pedro Carlos. "El pensamiento político de Antonio Cánovas del Castillo." In *Antonio Cánovas y el sistema político de la Restauración*, edited by Javier Tusell and Florentino Portero, 45–88. Madrid: Biblioteca Nueva, 1998.

González Herrán, José Manuel. "Reescritura en algunas crónicas periodísticas de Emilia Pardo Bazán (1912–1915)." In *Escritoras españolas en los medios de prensa: 1868–1936*, edited by Carmen Servén and Ivana Rota, 117–137. Sevilla: Renacimiento, 2014.

González Herrán, José Manuel, and Cristina Patiño Eirín. *Emilia Pardo Bazán: El periodismo*. A Coruña: Real Academia Galega, 2007.

Goode, Joshua. *Impurity of Blood: Defining Race in Spain, 1870–1930*. Baton Rouge: Louisiana State University Press, 2009.

Gortázar, Guillermo. *Alfonso XIII, hombre de negocios: Persistencia del Antiguo Régimen, modernización económica y crisis política, 1902–1931*. Madrid: Alianza, 1986.

Hacking, Ian. *The Taming of Chance*. Cambridge: Cambridge University Press, 1990

Harré, Rom. "Material Objects in Social Worlds." *Theory, Culture & Society* 19, no. 5–6 (2002): 23–33.

Harrison, Joseph. "The Regenerationist Movement in Spain after the Disaster of 1898." *European History Quarterly* 9, no. 1 (1979): 1–27.

Harrison, Joseph, and Alan Hoyle, eds. *Spain's 1898 Crisis: Regenerationism, Modernism, Post-Colonialism*. Manchester: Manchester University Press, 2000.

Hayles, N. Katherine. *Chaos Bound: Orderly Disorder in Contemporary Literature and Science*. Ithaca: Cornell University Press, 1990.

———. *The Cosmic Web: Scientific Field Models and Literary Strategies in the Twentieth Century*. Ithaca: Cornell University Press, 1984.

Henderson, Andrea. "Math for Math's Sake: Non-Euclidean Geometry, Aestheticism, and *Flatland*." *PMLA* 124, no. 2 (2009): 455–471.

Hobsbawm, Eric J. *Industry and Empire, from 1750 to the Present Day*. New York: The New Press, 1999.

Holguin, Sandie Eleanor. *Creating Spaniards: Culture and National Identity in Republican Spain*. Madison: University of Wisconsin Press, 2002.

Hormigón Blánquez, Mariano. "*El Progreso Matemático* (1891–1900): Un estudio sobre la primera revista matemática española." *Llull: Revista de la Sociedad Española de Historia de las Ciencias y de las Técnicas* 4, no. 6–7 (1981): 87–116.

Huelin, Emilio. "Petroleo cual agente de bárbaros incendios y de otras distintas maneras considerado." *Revista de España* 31, no. 131 (1873): 5–25.

Instituto Español de Comercio Exterior (ICEX). "New Technologies from Spain." http://icex.technologyreview.com/ (accessed February 1, 2021).

Jiménez Morales, María Isabel. "Emilia Pardo Bazán, cronista en París (1889)." *Revista de Literatura* 52, no. 140 (2008): 507–532.

Johnson, Roberta. *Crossfire: Philosophy and the Novel in Spain, 1900–1934*. Lexington: University Press of Kentucky, 1993.

Jones, C. A. "Galdós's *Marianela* and the Approach to Reality." *Modern Language Review* 56, no. 4 (1961): 515–519.

Julivert Casagualda, Manuel. *Una historia de la geología en España: En su contexto socioeconómico, cultural y político, y en el marco de la geología internacional.* Barcelona: Publicacions i Edicions de la Universitat de Barcelona, 2014.

Kern, Stephen. *The Culture of Time and Space, 1880–1918.* Cambridge: Harvard University Press, 2003.

Ketabgian, Tamara. *The Lives of Machines: The Industrial Imaginary in Victorian Literature and Culture.* Ann Arbor: University of Michigan Press, 2011.

Labanyi, Jo. *Gender and Modernization in the Spanish Realist Novel.* Oxford: Oxford University Press, 2000.

Lafuente, Antonio, and Tiago Saraiva. "The Urban Scale of Science and the Enlargement of Madrid (1851–1936)." *Social Studies of Science* 34, no. 4 (2004): 531–569.

Lakoff, George, and Mark Johnson. *Metaphors We Live By.* Chicago: University of Chicago Press, 1980.

Landry, Travis. *Subversive Seduction: Darwin, Sexual Selection, and the Spanish Novel.* Seattle: University of Washington Press, 2012.

Lara Ramos, Antonio. *Pedro Antonio de Alarcón.* Granada: Comares, 2001.

Lawless, Geraldine. *Modernity's Metonyms: Figuring Time in Nineteenth-Century Spanish Stories.* Lewisburg: Bucknell University Press, 2011.

León Roca, José Luis. *Vicente Blasco Ibáñez.* Valencia: Ajuntament de Valencia, 1997.

Li, Gloria Elsa. "Literatura y sociedad en *El intruso* y *La bodega* de Vicente Blasco Ibáñez." PhD diss., Stanford University, 1995. ProQuest (304216538).

Lieberman, Sima. *The Contemporary Spanish Economy: A Historical Perspective.* London: Routledge, 2006.

Lipp, Solomon. *Francisco Giner de los Ríos: A Spanish Socrates.* Waterloo: Wilfrid Laurier University Press, 1985.

Lissorgues, Yves, and Gonzalo Sobejano, eds. *Pensamiento y literatura en España en el siglo XIX: Idealismo, positivismo, espiritualismo.* Toulouse: Presses Universitaires du Mirail, 1998.

Litvak, Lily. "Abolición del tiempo y el espacio. El tren a fines del siglo XIX." *Compás de Letras. Monografías de Literatura Española,* no. 7 (1995): 239–254.

———. *Transformación industrial y literatura en España (1895–1905).* Madrid: Taurus, 1980.

Lope, Hans-Joachim. "Locomotoras: La poesía ferroviaria del siglo XIX. Aproximaciones hispano-alemanas." In *Poesía lírica y progreso tecnológico (1868–1939),* edited by Sabine Schmitz, 109–142. Madrid: Iberoamericana, 2003.

López, Ignacio-Javier. *Galdós y el arte de la prosa.* Barcelona: Promociones y Publicaciones Universitarias, 1993.

———. "Ortega Munilla y la doble génesis de *La desheredada.*" *Anales Galdosianos* 20, no. 2 (1985): 7–17.

López Catalán, Julián. *La industria y el comercio al alcance de los niños.* Barcelona: J. Bastinos e Hijo, 1874.

López Cerezo, José A., and José M. Sánchez Ron. *Ciencia, tecnología, sociedad y cultura en el cambio de siglo.* Madrid: Biblioteca Nueva, 2001.

López-Morell, Miguel Ángel. "Salamanca y la construcción del ferrocarril de Aranjuez." In *Ferrocarril y Madrid: Historia de un progreso,* edited by Francisco Polo Muriel,

María Jesús Matilla Quiza, and Manuel Benegas Capote, 13–44. Madrid: Fundación de los Ferrocarriles Españoles, 2002.

López-Morillas, Juan. *El krausismo español: Perfil de una aventura intelectual*. México: Fondo de Cultura Económica, 1980.

López-Ocón, Leoncio. *Breve historia de la ciencia española*. Madrid: Alianza, 2003.

López Piñero, José María. "Introducción." In *La Ciencia en la España del siglo XIX*, edited by José María López Piñero and Mariano Peset Reig, 11–18. Madrid: Marcial Pons, 1992.

López Piñero, José María, and Mariano Peset Reig, eds. *La Ciencia en la España del siglo XIX*. Madrid: Marcial Pons, 1992.

López Serrano, Juan. *La legalidad común: Solución política*. Madrid: Imprenta y litografía de Nicolás González, 1875.

Lozano Carbayo, Pilar. *El libro del tren: 150 años de ferrocarril en España*. Madrid: Guillomía Comunicación Gráfica, 1988.

Macías Picavea, Ricardo. *El problema nacional: Hechos, causas, remedios*. Madrid: Librería general de Victoriano Suárez, 1899.

Maeztu, Ramiro de. "El socialismo bilbaíno." *Germinal*, Julio 16, 1897.

———. "La obra del odio." *El Heraldo de Madrid*, Octubre 28, 1903.

Mallada, Lucas. *Los males de la patria y la futura revolución española: Consideraciones generales acerca de sus causas y efectos*. Edited by Francisco J. Flores Arroyuelo. Madrid: Alianza, 1969.

———. "Los progresos de la Geología en España durante el siglo XIX." In *Discursos leídos ante la Real Academia de Ciencias Exactas, Físicas y Naturales en la recepción pública del Sr. D. Lucas Mallada y Pueyo el día 29 de junio de 1897*, 7–66. Madrid: Imprenta de Luis Aguado, 1897.

———. *Proyecto de una nueva división territorial de España*. Madrid: Imprenta y estereotipia de El Liberal, 1881.

Maluquer de Motes, Jordi. "Factores y condicionamientos del proceso de industrialización en el siglo XIX: El caso español." In *La industrialización del norte de España*, edited by Jordi Maluquer de Motes, Emiliano Fernández de Pinedo, and José Luis Hernández Marco, 13–36. Barcelona: Crítica, 1988.

Maluquer de Motes, Jordi, Emiliano Fernández de Pinedo, and José Luis Hernández Marco, eds. *La industrialización del norte de España*. Barcelona: Crítica, 1988.

Marx, Karl. "Estranged Labor." In *Princeton Readings in Political Thought: Essential Texts Since Plato*, edited by Mitchell Cohen and Nicole Fermon, 438–448. Princeton: Princeton University Press, 1996.

Menéndez Onrubia, Carmen, and Julián Ávila Arellano. *El neorromanticismo español y su época: Epistolario de José Echegaray a María Guerrero*. Madrid: Consejo Superior de Investigaciones Científicas, 1987.

Mercer, Leigh. *Urbanism and Urbanity: The Spanish Bourgeois Novel and Contemporary Customs (1845–1925)*. Lewisburg: Bucknell University Press, 2013.

Mesonero Romanos, Ramón de. *Recuerdos de viaje por Francia y Bélgica*. Madrid: Oficinas de La Ilustración Española y Americana, 1881.

Ministerio de Fomento de España. "Líneas de alta velocidad." http://www.adif.es/ (accessed February 1, 2021).

Mirowski, Philip. *More Heat than Light: Economics as Social Physics, Physics as Nature's Economics*. Cambridge: Cambridge University Press, 1989.

Montero García, Feliciano. *El primer catolicismo social y la Rerum Novarum (1899–1902).* Madrid: Consejo Superior de Investigaciones Científicas, 1983.

Montes Doncel, Rosa Eugenia. "Variaciones de un tema novelesco: *Doña Perfecta* de Galdós y *El intruso* de Blasco Ibáñez." *Revista Hispánica Moderna* 58, no. 1–2 (2005): 39–59.

Muñoz Cortés, Manuel. "Variedad regional, lengua vernácula y conflicto lingüístico en el Bilbao del siglo XIX y su función en *El intruso* de Blasco Ibáñez." In *Estudis en memòria del professor Manuel Sanchis Guarner*, 215–224. Vol. 2. Valencia: Universitat de València, 1984.

Myers, Greg. "Nineteenth-Century Popularizations of Thermodynamics and the Rhetoric of Social Prophecy." *Victorian Studies* 29, no. 1 (1985): 35–66.

Nadal, Jordi. *El fracaso de la revolución industrial en España, 1814–1913.* Barcelona: Ariel, 1977.

Nieto-Galan, Agustí. "The Images of Science in Modern Spain: Rethinking the 'Polémica.'" In *The Sciences in the European Periphery during the Enlightenment*, edited by Kostas Gavroglu, 73–94. Dordrecht: Springer, 1999.

———. "A Republican Natural History in Spain around 1900: Odón de Buen (1863–1945) and His Audiences." *Historical Studies in the Natural Sciences* 42, no. 3 (2012): 159–189.

Núñez de Arce, Gaspar. "Causa de la precipitada decadencia y total ruina de la literatura nacional bajo los últimos reinados de la casa de Austria." In *Discursos leídos ante la Real Academia Española en la pública recepción del Excmo. señor Don Gaspar Núñez de Arce el día 21 de mayo de 1876*, 5–44. Madrid: Imprenta de T. Fortanet, 1876.

Olson, Richard. *Science and Scientism in 19th-Century Europe.* Urbana: University of Illinois Press, 2008.

Ortega Munilla, José. *El tren directo, relación contemporánea.* Madrid: Oficinas de La Ilustración Española y Americana, 1880.

Ortiz Villajos, José María. *Tecnología y desarrollo económico en la historia contemporánea: Estudio de las patentes registradas en España entre 1882 y 1935.* Madrid: Oficina Española de Patentes, 1999.

Otis, Laura. "The Metaphoric Circuit: Organic and Technological Communication in the Nineteenth Century." *Journal of the History of Ideas* 63, no. 1 (2002): 105–128.

———. *Networking: Communicating with Bodies and Machines in the Nineteenth Century.* Ann Arbor: University of Michigan Press, 2001.

———. *Organic Memory: History and the Body in the Late Nineteenth and Early Twentieth Centuries.* Lincoln: University of Nebraska Press, 1994.

———. "Science and Signification in the Early Writings of Emilia Pardo Bazán." *Revista de Estudios Hispánicos* 29, no. 1 (1995): 73–106.

Palacio Valdés, Armando. *La aldea perdida: Novela-poema de costumbres campesinas.* Madrid: Imprenta de los Hijos de M. G. Hernández, 1903.

———. *La espuma.* Barcelona: Imprenta de Henrich y Compañía en comandita, 1890.

Palomo Vázquez, Pilar, Pilar Vega Rodríguez, and Concepción Núñez Rey, eds. *Emilia Pardo Bazán, Periodista.* Madrid: Arco Libros, 2015.

Papanelopoulou, Faidra, Agustí Nieto-Galán, and Enrique Perdiguero, eds. *Popularizing Science and Technology in the European Periphery, 1800–2000.* Burlington: Ashgate, 2009.

Pardo Bazán, Emilia. "La ciencia amena." 11 parts. *Revista Compostelana* 1, no. 3 (1876): 17–18; 1, no. 4 (1876): 25–27; 1, no. 6 (1876): 41–43; 1, no. 7 (1876): 49–51; 1, no. 8

(1876): 57–59; 1, no. 9 (1876): 65–67; 1, no. 10 (1876): 73–75; 1, no. 11 (1876): 81–83; 1, no. 12 (1876): 89–91; 2, no. 13 (1876): 97–99; 2, no. 14 (1876): 105–107.

———. *La cuestión palpitante*. Madrid: Imprenta Central, 1883.

———. "La vida contemporánea." *La Ilustración Artística* 17, no. 875 (Octubre 3, 1898): 634.

———. "La vida contemporánea." *La Ilustración Artística* 18, no. 929 (Octubre 16, 1899): 666.

———. "La vida contemporánea." *La Ilustración Artística* 19, no. 958 (Mayo 7, 1900): 298.

———. "La vida contemporánea." *La Ilustración Artística* 20, no. 1031 (Septiembre 30, 1901): 634.

———. "La vida contemporánea." *La Ilustración Artística* 33, no. 1697 (Julio 6, 1914): 446.

———. *Obras completas*. Vol. 19, *Al pie de la Torre Eiffel*. Madrid: Establecimiento tipográfico de Idamor Moreno, 1899.

———. "Reflexiones científicas contra el darwinismo." 4 parts. *La Ciencia Cristiana*, no. 4 (1877): 289–298; no. 5 (1877): 18–55, 393–410, 481–495.

———. "Sud-Exprés." In *Obras completas*, 5–11. Vol. 36. Madrid: Pueyo, 1900.

Pérez de Perceval Verde, Miguel Á., and Alejandro Sánchez Rodríguez. "El capital extranjero en el desarrollo de la minería española del siglo XIX." In *La inversión extranjera en la minería española*, edited by Albert Broder et al., 7–69. Madrid: Instituto Geológico y Minero de España, 2014.

Pérez Galdós, Benito. *De vuelta a Italia*. Buenos Aires: Jesús Menéndez, 1900.

———. *Doña Perfecta*. Madrid: Imprenta y litografía de La Guirnalda, 1876.

———. *Gloria*. Madrid: Imprenta de José María Pérez, 1877

———. *La familia de León Roch*. Madrid: Imprenta y litografía de La Guirnalda, 1879.

———. "La novela en el tranvía." In *13 cuentos*, edited by Esteban Gutiérrez Díaz-Bernardo, 171–201. Madrid: Edaf, 2001.

———. *Marianela*. Madrid: Imprenta y litografía de La Guirnalda, 1878.

Pérez-Galdós, Enrique, and Natalia Pérez-Galdós. "La dualidad Echegaray." *Ingeniería y Territorio*, no. 79 (2007): 4–11.

Pérez Romero, Emilia. *El periodismo de Emilia Pardo Bazán*. Vigo: Editorial Academia del Hispanismo, 2016.

Peset, Mariano, and José Luis Peset. "Las universidades españolas del siglo XIX y las ciencias." *Ayer*, no. 7 (1992): 19–49.

Pohl-Valero, Stefan. "The 'Circulation' of Energy: Thermodynamics, National Culture, and Social Progress in Spain, 1868–1890." In *Popularizing Science and Technology in the European Periphery, 1800–2000*, edited by Faidra Papanelopoulou, Agustí Nieto-Galán, and Enrique Perdiguero, 115–134. Burlington: Ashgate, 2009.

———. *Energía y cultura: Historia de la termodinámica en la España de la segunda mitad del siglo XIX*. Bogotá, D.C.: Pontificia Universidad Javeriana, 2011.

———. "La comunicación de la termodinámica. Física, cultura y poder en la España de la segunda mitad del siglo XIX." *Memoria y Sociedad* 13, no. 27 (2009): 121–141.

———. "Termodinámica, pensamiento social y biopolítica en la España de la Restauración." *Universitas Humanística*, no. 69 (2010): 33–58.

Polo Muriel, Francisco, María Jesús Matilla Quiza, and Manuel Benegas Capote, eds. *Ferrocarril y Madrid: Historia de un progreso*. Madrid: Fundación de los Ferrocarriles Españoles, 2002.

Ponce, Juan Carlos. *Literatura y ferrocarril en España: Aspectos socioliterarios del ferrocarril en España*. Madrid: Fundación de los Ferrocarriles Españoles, 1996.

Prados de la Escosura, Leandro. "Foreign Capital in 19th Century Spain's Investment Boom." *European Review of Economic History* 24, no. 2 (2020): 314–331.

Pratt, Dale J. "'Fundamental Truths' to 'Playthings of Science': Science, Technology, and Modernity in Pardo Bazán." In *Approaches to Teaching the Writings of Emilia Pardo Bazán*, edited by Margot Versteeg and Susan Walter, 71–77. New York: Modern Language Association of America, 2017.

———. *Signs of Science: Literature, Science, and Spanish Modernity since 1868*. West Lafayette: Purdue University Press, 2001.

Rábano, Isabel. *Los cimientos de la Geología: La Comisión del Mapa Geológico de España (1849–1910)*. Madrid: Instituto Geológico y Minero de España, 2015.

Rabinbach, Anson. *The Human Motor: Energy, Fatigue, and the Origins of Modernity*. New York: BasicBooks, 1990.

Ramón y Cajal, Santiago. "Fundamentos racionales y condiciones técnicas de la investigación biológica." In *Discursos leídos ante la Real Academia de Ciencias Exactas, Físicas y Naturales en la recepción pública del Sr. D. Santiago Ramón y Cajal el día 5 de diciembre de 1897*, 7–82. Madrid: Imprenta de Luis Aguado, 1897.

———. *Los tónicos de la voluntad: Reglas y consejos sobre investigación científica*. Buenos Aires: Espasa-Calpe, 1971.

Ramos Gorostiza, José Luis. "'Carácter nacional' y decadencia en el pensamiento español." *Biblio 3W. Revista Bibliográfica de Geografía y Ciencias Sociales* 15, no. 860 (2010). http://www.ub.es/geocrit/b3w-860.htm (accessed February 15, 2021).

Real Academia de la Historia. "José Rodríguez Mourelo." http://dbe.rah.es/biografias/25713/jose-rodriguez-mourelo (accessed February 17, 2021).

Reig, Ramiro. *Vicente Blasco Ibáñez*. Madrid: Espasa, 2002.

———. "Vicente Blasco Ibáñez (1867–1928): Promotor de rebeldías." In *Liberales, agitadores y conspiradores. Biografías heterodoxas del siglo XIX*, edited by Isabel Burdiel and Manuel Pérez Ledesma, 331–361. Madrid: Espasa, 2000.

Rey Pastor, Julio. "El progreso de España en las ciencias y el progreso de las ciencias en España." In *Discursos del quinto congreso de la Asociación Española para el Progreso de las Ciencias celebrado en Valladolid del 7 al 22 de octubre de 1915*, 11–25. Madrid: Imprenta de Eduardo Arias, 1915.

Ribas, Pedro. "Regeneracionismo: una relectura." In *El Regeneracionismo en España: Política, educación, ciencia y sociedad*, edited by Vicent Salavert Fabiani and Manuel Suárez Cortina, 47–80. Valencia: Universitat de València, 2007.

Ribbans, Geoffrey. *History and Fiction in Galdós's Narratives*. Oxford: Clarendon, 1993.

Rodríguez Carracido, José. *Discurso leído en la solemne primera adjudicación de la medalla de su nombre al Excmo. Sr. D. José Echegaray el día 16 de junio de 1907*. Madrid: Imprenta de La Gaceta de Madrid, 1907.

Rodríguez Mourelo, José. "Concepto de la energía. Curso de ciencias naturales: Tercera conferencia, 12 de enero de 1882." *Revista Contemporánea* 8, no. 40 (1882): 275–311.

———. "Discurso del Ilmo. Sr. D. José Rodríguez Mourelo." In *Discursos leídos ante la Real Academia de Ciencias Exactas, Físicas y Naturales en la recepción pública del Ilmo. Sr. D. José Rodríguez Mourelo el 24 de mayo de 1903*, 3–43. Madrid: Imprenta de Luis Aguado, 1903.

Rojas, Francisco. *Termodinámica: Su historia, sus aplicaciones y su importancia*. Barcelona: Establecimiento tipográfico de Luis Tasso, 1876.

Rojas-Auda, Elizabeth. "Concha Espina y su obra comprometida: 1888–1936." PhD diss., University of Wisconsin–Madison, 1997. ProQuest (304456096).

Ruiz, Mario E. "El idealismo platónico en *Marianela* de Galdós." *Hispania: A Journal Devoted to the Teaching of Spanish and Portuguese* 53, no. 4 (1970): 870–880.

Saavedra, Ángel de, Duque de Rivas. "Viajes al Vesubio." In *Prosas*, 299–310. Vol. 5 of *Obras completas*. Madrid: Imprenta de la Biblioteca Nueva, 1855.

Sala Catalá, José. *Ideología y ciencia biológica en España entre 1860 y 1881. La difusión de un paradigma*. Madrid: Consejo Superior de Investigaciones Científicas, 1987.

Salavert Fabiani, Vicent, and Manuel Suárez Cortina, eds. *El Regeneracionismo en España: Política, educación, ciencia y sociedad*. Valencia: Universitat de València, 2007.

Sánchez Agustí, María. *La educación española a finales del XIX: Una mirada a través del periódico republicano* La Libertad. Lleida: Milenio, 2002.

Sánchez Ron, José Manuel. *Cincel, martillo y piedra: Historia de la ciencia en España (siglos XIX y XX)*. Madrid: Taurus, 2000.

———. "Las ciencias físico-matemáticas en la España del siglo XIX." *Ayer*, no. 7 (1992): 51–84

Sanz del Río, Julián. "Cartas inéditas de don Julián Sanz del Río." *Revista Europea* 1, no. 3 (1874): 66–71.

Sarmiento Ramírez, Ismael. "Cultura y cultura material: Aproximaciones a los conceptos e inventario epistemológico." *Anales del Museo de América*, no. 15 (2007): 217–236.

Scharff, Robert C. *Comte after Positivism*. Cambridge: Cambridge University Press, 2002.

Schivelbusch, Wolfgang. *The Railway Journey: The Industrialization of Time and Space in the 19th Century*. Berkeley: University of California Press, 1986.

Schmidt, Ruth. *Ortega Munilla y sus novelas*. Madrid: Revista de Occidente, 1973.

Schmitz, Sabine, ed. *Poesía lírica y progreso tecnológico (1868–1939)*. Madrid: Iberoamericana, 2003.

Schnepf, Michael A. "A Note on Galdós, Ortega Munilla, and *La desheredada*." *Romance Notes* 39, no. 1 (1998): 3–7.

Serrano Fatigati, Enrique. "La evolución en la naturaleza." *Revista de la Universidad de Madrid*, no. 3 (1874): 292–317, 496–512.

Serres, Michel. *Hermes IV: La Distribution*. Paris: Minuit, 1977.

———. *Hermes—Literature, Science, Philosophy*. Edited by David F. Bell and Josué V. Harari. Baltimore: Johns Hopkins University Press, 1982.

Servén, Carmen, and Ivana Rota, eds. *Escritoras españolas en los medios de prensa: 1868–1936*. Sevilla: Renacimiento, 2014.

Shoemaker, William Hutchinson. *The Novelistic Art of Galdós*. 3 vols. Valencia: Albatros Hispanófila, 1980–1982.

Sieburth, Stephanie Anne. *Inventing High and Low: Literature, Mass Culture, and Uneven Modernity in Spain*. Durham: Duke University Press, 1994.

Sierra, Sarah. "Time and the Environment: 'Slow Violence' in Benito Pérez Galdós's *Doña Perfecta* and *Marianela*." *Confluencia: Revista Hispánica de Cultura y Literatura* 35, no. 1 (2019): 41–56.

Simmel, Georg. *Simmel on Culture: Selected Writings*. Edited by David Frisby and Mike Featherstone. London: Sage, 2000.

Smith, Jonathan. *Fact and Feeling: Baconian Science and the Nineteenth-Century Literary Imagination*. Madison: University of Wisconsin Press, 1994.

Smith, Paul. *Vicente Blasco Ibáñez: Una nueva introducción a su vida y obra*. Santiago de Chile: Andrés Bello, 1972.

Sosa-Velasco, Alfredo. *Médicos escritores en España 1885–1955*. Woodbridge: Tamesis, 2010.

Stowe, William W. *Going Abroad: European Travel in Nineteenth-Century American Culture*. Princeton: Princeton University Press, 1994.

Sultzbach, Kelly Elizabeth. *Ecocriticism in the Modernist Imagination: Forster, Woolf, and Auden*. Cambridge: Cambridge University Press, 2016.

Taylor, Peter J. *Political Geography: World-Economy, Nation-State and Locality*. London: Longman, 1985.

Tilley, Christopher Y. *Metaphor and Material Culture*. Oxford: Blackwell, 2000.

Torres, Pablo. "Político, periodista, escritor . . . Blasco Ibáñez, un personaje incómodo." *Carta de España*, no. 737 (2017): 32–33.

Torres Muñoz de Luna, Ramón. "Las dos Isabeles: Composición poética para solemnizar la inauguración del ferro-carril de León a Valencia." In Biblioteca Nacional, Madrid, MSS/12944/103 (1863).

Tortella Casares, Gabriel. *El desarrollo de la España contemporánea: Historia económica de los siglos XIX y XX*. Madrid: Alianza, 1998.

Tsanoff, Radoslav Andrea. *Civilization and Progress*. Lexington: University Press of Kentucky, 1971.

Tsuchiya, Akiko. *Images of the Sign: Semiotic Consciousness in the Novels of Benito Pérez Galdós*. Columbia: University of Missouri Press, 1990.

Tusell, Javier, and Florentino Portero, eds. *Antonio Cánovas y el sistema político de la Restauración*. Madrid: Biblioteca Nueva, 1998.

Unamuno, Miguel de. *Por tierras de Portugal y de España*. Madrid: Renacimiento, 1911.

———. *Vida de Don Quijote y Sancho*. Madrid: Librería de Fernando Fe, 1905.

Underwood, Ted. *The Work of the Sun: Literature, Science, and Political Economy, 1760–1860*. New York: Palgrave Macmillan, 2005.

Universitat de València, ed. *Estudis en memòria del professor Manuel Sanchis Guarner*. 2 vols. Valencia: Universitat de València, 1984.

Uriarte Ayo, Rafael. "La minería vizcaína del hierro en las primeras etapas de la industrialización." In *La industrialización del norte de España*, edited by Jordi Maluquer de Motes, Emiliano Fernández de Pinedo, and José Luis Hernández Marco, 154–177. Barcelona: Crítica, 1988.

Valle, Ángela del. *Aportación bio-bibliográfica a la historia de la ciencia: Universidad Central 1886–1902*. Madrid: Narcea, 1998.

Vallejo, Catharina. "Emilia Pardo Bazán, Gender, Modernity and Nationalism at the Paris World Exhibitions of 1889 and 1900." *Revista Canadiense de Estudios Hispánicos* 32, no. 3 (2008): 453–473.

Versteeg, Margot, and Susan Walter, eds. *Approaches to Teaching the Writings of Emilia Pardo Bazán*. New York: Modern Language Association of America, 2017.

Vicens i Vives, Jaume. *Manual de historia económica de España*. Barcelona: Teide, 1959.

Vicuña, Gumersindo. *Discurso leído en la Universidad Central en el acto de apertura del curso académico de 1875 a 1876*. Madrid: Imprenta de José M. Ducazcal, 1875.

———. *Elementos de física al alcance de todo el mundo*. Madrid: Noguera, 1874.

———. *Manual de física popular*. Madrid: Tipografía de Estrada, 1878.

Viñes Millet, Cristina. "El africanismo de Pedro Antonio de Alarcón." In *Pedro Antonio de Alarcón y la guerra de África: Del entusiasmo romántico a la compulsión colonial*, edited by José Antonio González Alcantud and Manuel Lorente Rivas, 45–60. Barcelona: Anthropos, 2004.

Woodward, Ian. *Understanding Material Culture*. London: Sage, 2007.

Yáñez, María Paz. "El dilema discursivo en *Marianela*." *Anales Galdosianos*, no. 29–30 (1994–95): 51–62.

Yengoyan, Aram A. Review of *Simmel on Culture: Selected Writings*, by David Frisby and Mike Featherstone, eds. *Comparative Studies in Society and History* 44, no. 3 (2002): 620–625.

Index

About the Author

Óscar Iván Useche received his Ph.D. in Latin American and Iberian cultures from Columbia University and is currently an associate professor of Spanish at Ursinus College. He also holds degrees in engineering and literature. As a specialist in modern peninsular studies, his research and teaching focus on exploring the interaction between industrial modernization and cultural production in *fin-de-siglo* Spain. His work has appeared in a variety of academic publications, including the *Journal of Spanish Cultural Studies*, *Decimonónica*, and *Siglodiecinueve*. He is presently at work on a new book project about failed inventions, scientific deception, and literary forgery at the turn of the twentieth century.